AF822601

Regulation of Carbohydrate Metabolism

Volume I

Editor

Rivka Beitner, Ph.D.
Associate Professor of Biochemistry
Department of Life Sciences
Bar-Ilan University, Ramat-Gan
Israel

CRC Press
Taylor & Francis Group
Boca Raton London New York

CRC Press is an imprint of the
Taylor & Francis Group, an **informa** business

First published 1985 by CRC Press
Taylor & Francis Group
6000 Broken Sound Parkway NW, Suite 300
Boca Raton, FL 33487-2742

Reissued 2018 by CRC Press

© 1985 by Taylor & Francis
CRC Press is an imprint of Taylor & Francis Group, an Informa business

No claim to original U.S. Government works

This book contains information obtained from authentic and highly regarded sources. Reasonable efforts have been made to publish reliable data and information, but the author and publisher cannot assume responsibility for the validity of all materials or the consequences of their use. The authors and publishers have attempted to trace the copyright holders of all material reproduced in this publication and apologize to copyright holders if permission to publish in this form has not been obtained. If any copyright material has not been acknowledged please write and let us know so we may rectify in any future reprint.

Except as permitted under U.S. Copyright Law, no part of this book may be reprinted, reproduced, transmitted, or utilized in any form by any electronic, mechanical, or other means, now known or hereafter invented, including photocopying, microfilming, and recording, or in any information storage or retrieval system, without written permission from the publishers.

For permission to photocopy or use material electronically from this work, please access www. copyright.com (http://www.copyright.com/) or contact the Copyright Clearance Center, Inc. (CCC), 222 Rosewood Drive, Danvers, MA 01923, 978-750-8400. CCC is a not-for-profit organiza-tion that provides licenses and registration for a variety of users. For organizations that have been granted a photocopy license by the CCC, a separate system of payment has been arranged.

Trademark Notice: Product or corporate names may be trademarks or registered trademarks, and are used only for identification and explanation without intent to infringe.

A Library of Congress record exists under LC control number: 84012060

Publisher's Note
The publisher has gone to great lengths to ensure the quality of this reprint but points out that some imperfections in the original copies may be apparent.

Disclaimer
The publisher has made every effort to trace copyright holders and welcomes correspondence from those they have been unable to contact.

ISBN 13: 978-1-315-89716-5 (hbk)
ISBN 13: 978-1-351-07626-5 (ebk)

Visit the Taylor & Francis Web site at http://www.taylorandfrancis.com and the
CRC Press Web site at http://www.crcpress.com

THE EDITOR

Rivka Beitner, Ph.D., is Associate Professor of Biochemistry in the Department of Life Sciences at Bar-Ilan University at Ramat-Gan, Israel.

Dr. Beitner obtained her training at Bar-Ilan University, Ramat-Gan, Israel, receiving the B.Sc. degree in 1961 and the M.Sc. degree in 1963. She received the Ph.D. degree in 1970 from McGill University, Montreal, Canada. She served as a Research Fellow at the Lady Davis Institute for Medical Research of the Jewish General Hospital and McGill University, Montreal, Canada, from 1967 to 1970 and as a Lecturer, Assistant Professor and Associate Professor of Biochemistry at Bar-Ilan University from 1970 to the present. It was in 1979 that she assumed her present position.

Dr. Beitner is a member of the Israeli Society for Biochemistry, the Israeli Society for Endocrinology, and the Israeli Society for Diabetes. She has been the recipient of many research grants from the Muscular Dystrophy Association of America; the U.S.-Israel Binational Science Foundation; the Medical Research Council of Canada; the Ministry of Commerce and Industry, Israel, and the Research Authority, Bar-Ilan University.

Dr. Beitner is the author of more than 70 papers and abstracts. Her current major research interests relate to the regulation of carbohydrate metabolism.

CONTRIBUTORS

Ramon Bartrons, Ph.D.
Department of Biochemistry
Faculty of Medicine
University of Barcelona
Barcelona, Spain

Nava Bashan, Ph.D.
Pediatric Research Laboratory
Soroka Medical Center
Faculty of Health Sciences
Ben Gurion University of the Negev
Beer Sheva, Israel

Rivka Beitner, Ph.D.
Associate Professor of Biochemistry
Department of Life Sciences
Bar-Ilan University
Ramat-Gan, Israel

Frank Clarke, Ph.D.
Lecturer
Griffith University
School of Science
Nathan, Australia

Valentine A. Duruibe, M.S.
Graduate Research Associate
Department of Pharmacology
College of Medicine
The Ohio State University
Columbus, Ohio

Erich Eigenbrodt, Prof. Dr.
Institut fur Biochemie und Endokrinologie
Fachbereich Veterinarmedizin und Tierzucht
Justus-Liebig-Universitat Giessen
Giessen, West Germany

P. Fister, Ph.D.
Medizinisch Wissenschaftliche Abteilung
Biotest Pharma
Frankfurt, West Germany

Alisa Gutman, M.D.
Professor of Biochemistry
Department of Biochemistry
Hadassah University Hospital and
Hebrew University-Hadassah Medical School
Jerusalem, Israel

Hans Werner Hofer, Prof. Dr.
University of Konstanz
Faculty of Biology
Konstanz, West Germany

Louis Hue, M.D., Ph.D.
Hormone and Metabolic Research Unit
International Institute of Molecular and Cellular Pathology
University of Louvain Medical School
Brussels, Belgium

Norman Kalant, Ph.D.
Lady Davis Institute for Medical Research
Sir Mortimer B. Davis-Jewish General Hospital and
Department of Medicine,
McGill University
Montreal, Quebec, Canada

Don Morton, Ph.D.
Principal Research Scientist
Meat Research Laboratories
CSIRO Division of Food Research
Cannon Hill, Australia

Shimon W. Moses, M.D., Ph.D.
Professor of Pediatrics
Pediatric Research Laboratory
Soroka Medical Center
Faculty of Health Sciences
Ben Gurion University of the Negev
Beer Sheva, Israel

M. Reinacher, Prof. Dr.
Institut fur Veterinar-Pathologie
Fachbereich Veterinarmedizin und Tierzucht
Justus-Liebig-Universtat Giessen
Giessen, West Germany

G. Rijksen, Ph.D.
Department of Hematology
Division of Medical Enzymology
Academic Hospital
Utrecht, The Netherlands

Eleazar Shafrir, Ph.D.
Professor and Head
Department of Clinical Biochemistry
Hadassah University Hospital and
Hebrew University-Hadassah Medical School
Jerusalem, Israel

Alfred E. Slonim, M.D.
Pediatric Department
North Shore University Hospital
Manhasset, New York

G. E. J. Staal, Ph.D.
Professor in Medical Enzymology
Department of Hematology
Academic Hospital
Utrecht, The Netherlands

Petra Stephan, M.S.
Postgraduate Student
Griffith University
School of Science
Nathan, Australia

Gopi A. Tejwani, Ph.D.
Assistant Professor
Department of Pharmacology
The Ohio State University
College of Medicine
Columbus, Ohio

John Weidemann, B.Sc.
Scientist
Meat Research Laboratory
CSIRO Division of Food Research
Cannon Hill, Australia

John E. Wilson, Ph.D.
Professor
Department of Biochemistry and
The Neuroscience Program
Michigan State University
East Lansing, Michigan

TABLE OF CONTENTS

Volume I

Volume II

Chapter 1

GLUCOSE-1,6-BISPHOSPHATE — THE REGULATOR OF CARBOHYDRATE METABOLISM

Rivka Beitner

TABLE OF CONTENTS

I. INTRODUCTION

Experiments in our as well as in other laboratories during the past few years have led us to realize that glucose-1,6-bisphosphate (Glc-1,6-P_2) is a powerful regulator of carbohydrate metabolism (for reviews see References 1 and 1a). The unique feature of Glc-1,6-P_2 is that it acts as a powerful regulator common to several key enzymes in the different pathways of carbohydrate metabolism. This compound is not a biosynthetic precursor or intermediate in a metabolic pathway, but acts as an intracellular signal, its concentration being controlled by different physiological, hormonal, and pathological conditions, leading to concomitant changes in the activities of the key enzymes which are modulated by this regulator. By this action Glc-1,6-P_2 controls the different pathways of carbohydrate metabolism under various conditions. The list of factors and conditions which affect cellular Glc-1,6-P_2 levels is growing fast.

II. REGULATION OF KEY ENZYMES OF CARBOHYDRATE METABOLISM BY GLC-1,6-P_2

Glc-1,6-P_2 acts as a powerful regulator common to serveral key enzymes in carbohydrate metabolism. Figure 1 shows the key enzymes whose activities are regulated by Glc-1,6-P_2. Glc-1,6-P_2 affects all the three key enzymes in glycolysis, acting as a potent activator (deinhibitor) of phosphofructokinase[2-6] (the rate-limiting enzyme in glycolysis), and pyruvate kinase,[7] and an inhibitor of hexokinase.[5,8-11] It is also an activator and deinhibitor of phosphoglucomutase,[8,12,13] and an inhibitor of fructose 1,6-bisphosphatase[14] (one of the key enzymes in gluconeogenesis), as well as an inhibitor of 6-phosphogluconate dehydrogenase,[15] which is one of the key enzymes in the pentose phosphate pathway. The regulatory action of Glc-1,6-P_2 on all these key enzymes suggests that this compound regulates the different pathways of glucose metabolism.

The activatory (deinhibitory) action of Glc-1,6-P_2 on phosphofructokinase activity was demonstrated in skeletal muscle,[2-4] brain,[3] red blood cells,[5,6] heart, and adipose tissue.[4]

Its inhibitory action of hexokinase activity was also demonstrated in many mammalian tissues,[5,8-11] such as muscle, red blood cells, adipose tissue, brain, skin, and Ehrlich tumor cells. In our early experiments we found a selective inhibition of hexokinase type II by Glc-1,6-P_2.[8] At that time, however, we were not aware of the influence of pH and Mg^{2+} concentration on the sensitivity of the inhibition by Glc-1,6-P_2.[10] Rose and Warms,[10] keeping the Mg^{2+} concentration low to minimize formation of Mg-α-Glc-1,6-P_2 which is not inhibitory, have shown that at pH 6 to 7 Glc-1,6-P_2 is a much better inhibitor of hexokinase than glucose 6-P, the well-known inhibitor of this enzyme. Upon reinvestigating effect of Glc-1,6-P_2 under these new conditions (low pH and low Mg^{2+}), we found inhibition not only of type II hexokinase, but also of type I (the predominant form in erythrocytes and brain). However, the affinity muscle type II hexokinase for this inhibitor is much greater. The concentration of Glc-1,6-P_2 in rat skeletal muscle is about tenfold lower than in erythrocytes or brain, and in this range of concentrations the type II hexokinase is selectively inhibited. In view of the different concentrations of the Glc-1,6-P_2 present in muscle and erythrocytes

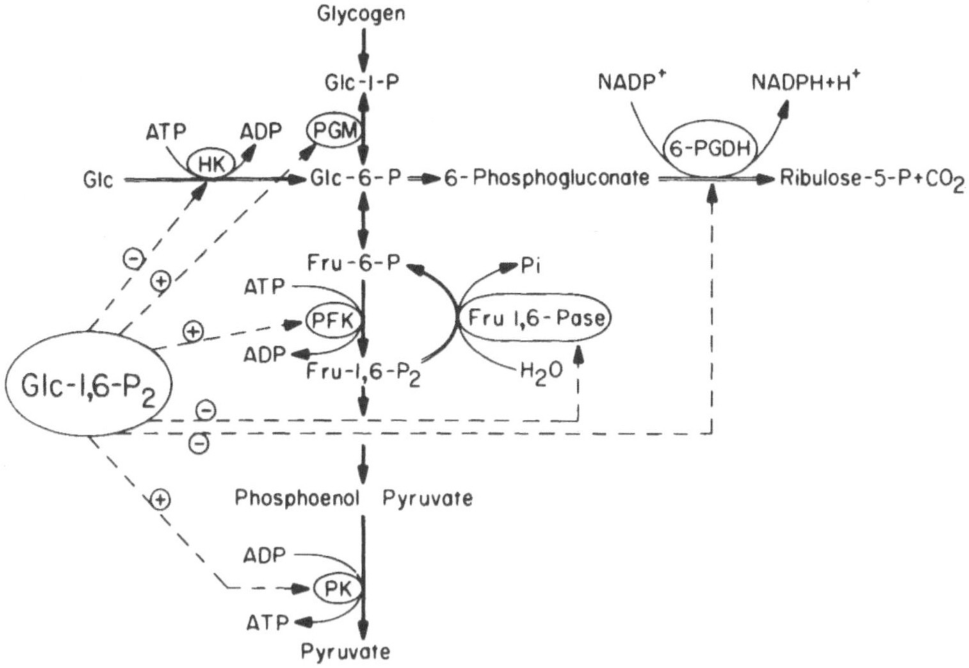

FIGURE 1. The regulatory action of Glc-1,6-P_2 on key enzymes of carbohydrate metabolism: HK, hexokinase; PFK, phosphofructokinase; PK, pyruvate kinase; PGM, phosphoglucomutase; 6-PGDH, 6-phosphogluconate dehydrogenase; and Fru 1,6-Pase, fructose 1,6-diphosphatase. ⊕, activation; ⊖, inhibition.

or brain and the different affinities of the two types of hexokinase to inhibition by Glc-1,6-P_2, it is probable that Glc-1,6-P_2 plays a regulator role in vivo on both muscle (type II), as well as on erythrocyte or brain (type I) hexokinase.

Glc-1,6-P_2 is also able to stimulate the pyruvate kinase from liver or red blood cells[7] by countering ATP inhibition.

Glc-1,6-P_2 was first shown by Leloir et al.[12] to be the coenzyme in the phosphoglucomutase reaction. Our experiments[8,13] have also shown that Glc-1,6-P_2 acts not only as a coenzyme in the phosphoglucomutase reaction, but also releases this enzyme from inhibition by ATP or citrate, similarly to its deinhibitory action on the activities of phosphofructokinase and pyruvate kinase. It has been suggested[16] that phosphoglucomutase (like phosphofructokinase or hexokinase) exists in state of submaximal efficiency in that native state of the cell. This enzyme has been categorized as a regulatory enzyme,[17] since intracellular substrate product ratio may be significantly displaced from equilibrium.[18,19] It has been suggested that in brain under certain conditions, e.g., following birth or during recovery from controlled anoxia, the phosphoglucomutase reaction may be a rate-limiting step in glycogen metabolism.[20]

It was reported that Glc-1,6-P_2 acts as an inhibitor of liver fructose 1,6-bisphosphatase[14] and an activator (deinhibitor) of liver and red blood cells pyruvate kinase.[7] We have shown that Glc-1,6-P_2 also exerts a potent inhibitory effect on the activity of 6-phosphogluconate dehydrogenase from yeast, muscle, liver, adipose tissue, and testis.[15] These findings suggest that Glc-1,6-P_2 may have a regulatory influence on the pentose phosphate pathway.

It has been recently reported[21] that Glc-1,6-P_2 also activates the inorganic pyrophosphate: D-fructose-6-phosphate 1-phosphotransferase from mung beans. We have found that Glc-1,6-P_2 is present in various plants.[123] It is possible therefore, that Glc-1,6-P_2 has a regulatory role not only in mammalian tissues but also in plants.

Table 1
EFFECTORS OF CELLULAR Glc-1,6-P_2 LEVELS

Effectors	Tissue	Changes in Glc-1,6-P_2 levels[a]	Ref.
Dibutyryl cAMP	Diaphragm	↑	13
	Cultured muscle	↑	23
Dibutyryl cGMP	Diaphragm	↓	24
	Cultured muscle	↓	23
Ca^{2+}-ionophore A23187	Diaphragm	↓	25
	Cultured muscle	↓	23
Epinephrine	Diaphragm	↑	13
	Cultured muscle	↑	23
Epinephrine + propranolol	Cultured muscle	↓	23
Vasopressin	Cultured muscle	↓	23
Serotonin	Skeletal muscle	↓	27, 28
	Skin	↓	27
Bradykinin	Skeletal muscle	↓	43
	Skin	↓	42
Muscular dystrophy	Skeletal muscle	↓	38, 39, 45, 61
Fasting	Skeletal muscle		
	Normal	↓	39, 61
	Dystrophic	↓	39, 61
Refeeding	Skeletal muscle		
	Normal	↑	39, 61
	Dystrophic	↑	39, 61
Phospholipase A_2	Diaphragm	↓	40
Lysolecithin	Diaphragm	↓	40
Anoxia	Diaphragm	↓	73
Ischemia	Brain	↓	74
Aerobiosis	Diaphragm	↑	73
Differentiation (fusion)	Cultured muscle	↑	23
Growth	Skeletal muscle	↑	62
	Heart	↔	62
	Brain	↓	80
	Skin	↓	116
Old age	Skeletal muscle	↓	82
Local anesthetics	Diaphragm	↓	41
Lithium	Skeletal muscle		
	Normal	↓	92, 93
	Dystrophic	↓	93
	Diaphragm	↓	92
	Brain	↓	91, 92
	Liver	↔	92
Trifluoperazine	Skeletal muscle	↑	94
Glucose	Pancreatic islets	↑	95

[a] ↑, increase; ↓, decrease; and ↔, no change.

III. CHANGES IN THE LEVELS OF GLC-1,6-P_2 AND IN THE ACTIVITIES OF KEY ENZYMES OF CARBOHYDRATE METABOLISM INDUCED BY VARIOUS FACTORS AND CONDITIONS

A long list of factors and conditions affect the intracellular levels of Glc-1,6-P_2, resulting in concomitant changes in the activities of the enzymes which are regulated by this modulator. The conditions which affect cellular Glc-1,6-P_2 levels are summarized in Table 1.

A. The Effect of Cyclic AMP, Cyclic GMP, and Calcium on Cellular Glc-1,6-P_2 Levels

It was found that the "second messenger triumvirate", namely, cyclic AMP, cyclic GMP, and calcium, control the cellular concentration of Glc-1,6-P_2. We found that dibutyryl cyclic AMP produced a significant increase in Glc-1,6-P_2 levels in the isolated rat diaphragm muscle.[13] This increase in the Glc-1,6-P_2 was accompanied by a concomitant activation of phosphofructokinase and phosphoglucomutase and a reduction in hexokinase activity.[13] An increase in Glc-1,6-P_2 concentration induced by dibutyryl cyclic AMP was also demonstrated by Wakelam and Pette in muscle cells in culture (fused myotubes).[22,23] The increase in Glc-1,6-P_2 in cultured muscle was accompanied by an increase in the fructose 1,6-bisphosphate/fructose 6-phosphate ratio, implying an activation of phosphofructokinase.[23]

Dibutyryl cyclic GMP exerted an opposite effect to that of dibutyryl cyclic AMP, namely, it decreased the cellular concentration of Glc-1,6-P_2 and reduced phosphofructokinase activity. This effect was demonstrated both in the isolated rat diaphragm muscle[24] and in cultured muscle tissue.[22,23]

Ca^{2+}-ionophore A23187 induced a similar effect to that of dibutyryl cyclic GMP. In the rat diaphragm muscle,[25] as well as in cultured muscle tissue,[22,23] this compound exerted a marked decrease in the levels of Glc-1,6-P_2. The decrease in Glc-1,6-P_2 was accompanied by a concomitant reduction in the activities of phosphofructokinase and phosphoglucomutase, when assayed under conditions in which these enzymes are sensitive to activation by Glc-1,6-P_2.[25] A decrease in the fructose 1,6-bisphosphate/fructose 6-phosphate ratio induced by the ionophore was also demonstrated in the cultured muscle.[23]

B. The Control of Glc-1,6-P_2 Levels by Hormonal Stimulation

As reviewed recently by Berridge,[26] most hormones and neurotransmitters exert their effects through two classes of receptors. One major class operates mainly through cyclic AMP, whereas another major class seems to act through the combined action of calcium and cyclic GMP. As mentioned above, cyclic AMP increases, whereas calcium and cyclic GMP decrease Glc-1,6-P_2 levels in muscle. It was found that hormones which act through cyclic AMP, exert an increase in Glc-1,6-P_2 concentration, whereas hormones or neurotransmitters which act through calcium and cyclic GMP, decrease Glc-1,6-P_2 level.

Epinephrine, which increases cyclic AMP concentration in muscle, was shown to increase the cellular Glc-1,6-P_2 concentration. This effect of epinephrine was demonstrated by us in the isolated rat diaphragm muscle[13] as well as by Wakelam and Pette in cultured muscle.[23] The rise in Glc-1,6-P_2 in diaphragm muscle induced by epinephrine resulted in a concomitant activation of phosphofructokinase and phosphoglucomutase and a reduction in hexokinase activity.[13] It was also shown that Glc-1,6-P_2 in the concentration range found after epinephrine inhibited the diaphragm hexokinase and activated (deinhibited) phosphofructokinase and phosphoglucomutase. These results suggest that Glc-1,6-P_2 may be a new second messenger of epinephrine action on the glycogenolytic and glycolytic processes; through the action of Glc-1,6-P_2 the activities of hexokinase, phosphoglucomutase, and phosphofructokinase may be synchronized with the cyclic AMP-mediated activation of glycogen phosphorylase to achieve an increase in total glycogenolysis and glycolysis and a concomitant reduction in glucose utilization by muscle.

An increase in Glc-1,6-P_2 and a concomitant activation of phosphofructokinase induced by epinephrine, was also demonstrated in muscle cells in culture (fused myotubes), as reflected by the increase in fructose 1,6-bisphosphate/fructose 6-phosphate ratio.[23]

On the other hand, addition of epinephrine + β-blocker, propranolol, resulted in a decrease in Glc-1,6-P_2 concentration in the cultured muscle, accompanied by a decrease in the fructose 1,6-bisphosphate/fructose 6-phosphate ratio.[23]

These results show that β-receptor stimulation (acting through cyclic AMP) induces an increase in Glc-1,6-P_2 and phosphofructokinase activity, whereas α-receptor stimulation

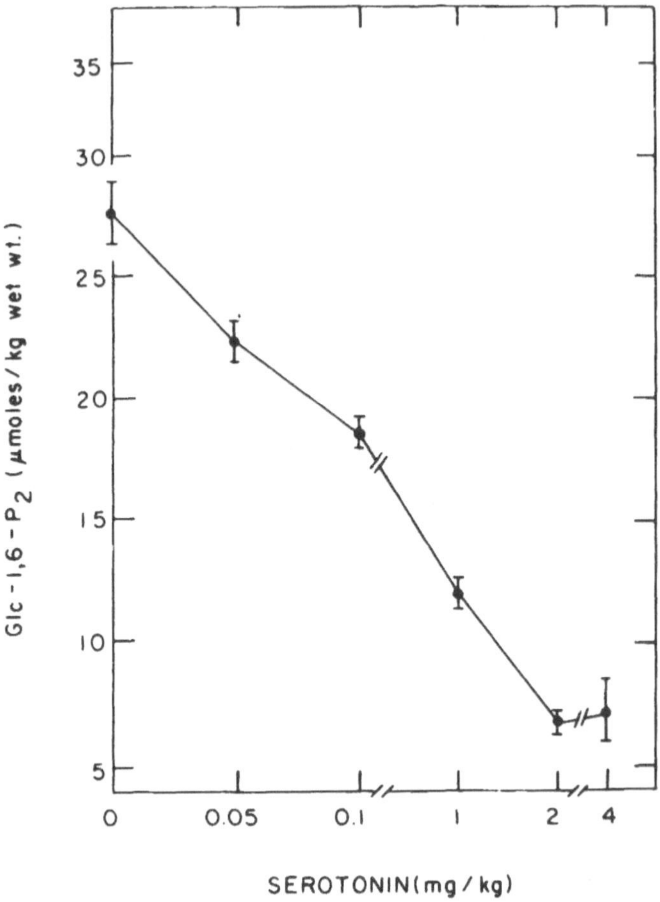

FIGURE 2. Dose response curve of the effect of serotonin on Glc-1,6-P_2 levels in the tibialis anterior muscle. Rats (180 to 220 g) were injected intramuscularly with different concentrations of serotonin for 20 min. Data are means ± SEM (represented by the bars) for 5 to 15 experiments. Each point is significantly different from controls (p <0.005). (From Beitner, R., Frucht, H., and Kaplansky, M., *Int. J. Biochem.*, 15, 935, 1983. With permission.)

(acting through calcium and cyclic GMP[26]) produces a decrease in Glc-1,6-P_2 and a concomitant inhibition of this enzyme.

Vasopressin, another hormone which is thought to produce both a rise in intracellular free Ca^{2+} concentration and a rise in cyclic GMP, also induced a decrease in Glc-1,6-P_2 concentration in cultured muscle, accompanied by a reduction in the fructose 1,6-bisphosphate/fructose 6-phosphate ratio.[23]

Recent experiments from our laboratory have shown that serotonin (5-hydroxytryptamine) affects cellular concentration of Glc-1,6-P_2.[27,28] Injection of serotonin induced a significant, dose-dependent decrease in the level of Glc-1,6-P_2 in rat skin[27] and in rat tibialis anterior muscle (Figure 2).[28] The decrease in Glc-1,6-P_2 concentration induced by serotonin was most probably mediated through calcium and cyclic GMP. Serotonin was shown to induce a significant increase in cyclic GMP concentration in the rat tibialis anterior muscle.[28] Like in other conditions, the changes in cellular Glc-1,6-P_2 concentration affected the activities of the enzymes which are modulated by Glc-1,6-P_2. It was found that the decrease in Glc-1,6-P_2 induced by serotonin was accompanied by a concomitant reduction in the activity of phosphofructokinase, when assayed under regulatory conditions in which the enzyme was subject to allosteric regulation.[28] This reduction was also reflected in a decrease in the concentration of fructose 1,6-bisphosphate, which is the product of phosphofructokinase reaction and also an activator of this enzyme (for reviews see References 29a, 29b, and 30). The activity of phosphoglucomutase was also markedly reduced by serotonin, as a result of the decrease in Glc-1,6-P_2, its potent activator.[28]

The serotonin-induced decrease in Glc-1,6-P_2 levels, mediating the decrease in phosphofructokinase activity and thereby in glycolysis, in muscle and skin, may be involved in the mechanism of the pathological influences of serotonin on these tissues. Serotonin was reported to induce muscle weakness and morphological changes in skeletal muscle, similar to the pathological changes occurring in muscular dystrophy.[31,32] In muscular dystrophy a defect in serotonin uptake into platelets was found,[33] and the blood serotonin levels of genetically dystrophic chickens were found to be significantly elevated.[34] Treatment of dystrophic chickens and hamsters with serotonin antagonists was reported to reduce considerably the elevated plasma creatine phosphokinase levels and to delay the symptoms of dystrophy.[34,35] It was also reported that in humans with carcinoid tumors which produce large amounts of circulating serotonin, a myopathy can develop, which may be corrected by serotonin antagonists.[36,37] Based on all these findings it has been suggested that serotonin plays a role in some diseases of muscle. In this regard it should be noted that a similar pathological decrease in Glc-1,6-P_2 level and phosphofructokinase activity to that induced by serotonin, we found in muscular dystrophy[38] as well as under any condition or treatment which leads to muscle weakness or damage.[25,39-41] Serotonin may also be involved in various skin abnormalities, e.g., those occurring in the carcinoid syndrome. As serotonin is one of the components of insect venoms, it may also be involved in the skin changes induced by the venoms.

Our recent experiments have revealed that bradykinin is another hormone which affects cellular Glc-1,6-P_2. The injection of this monopeptide hormone to rats induced a marked decrease in the levels of glc-1,6-P_2 in skin,[42] as well as in the tibialis anterior muscle.[43] The decrease in Glc-1,6-P_2 was accompanied by a concomitant decrease in the activities of phosphofructokinase and phosphoglucomutase.[43] Bradykinin also induced an increase in cyclic GMP levels in the tibialis anterior muscle.[43] A bradykinin-induced increase in cyclic GMP was also found in other tissues and was shown to be Ca^{2+} dependent.[44] It is possible therefore that cyclic GMP and Ca^{2+} may mediate the decrease in intracellular concentration of Glc-1,6P_2 induced by bradykinin. The bradykinin-induced decrease in cellular Glc-1,6-P_2 levels and the resulting reduction in glycolysis may be involved in the mechanism of the pathologic influences of this hormone in various clinical conditions.

C. The Regulatory Role of Glc-1,6-P_2 in Muscular Dystrophy

In muscle from genetically dystrophic mice of strain 129 ReJ, we found a striking decrease in the levels of Glc-1,6-P_2.[38,39,45] This reduction occurs in the early stages of the desease which suggests that these changes may be primary and not secondary in relation to the dystrophic process. The decrease in Glc-1,6-P_2 was accompanied by a marked reduction in the activities of phosphofructokinase and phosphoglucomutase and an enhancement of hexokinase and 6-phosphogluconate dehydrogenase. These changes in enzymes' activities, induced by the fall in Glc-1,6-P_2, reflect a shift from glycolysis to an enhanced glucose utilization through the pentose phosphate pathway, and thus may account for the increase in lipid (mediated by increased provision of NADPH), and RNA and DNA content (mediated by increased provision of pentoses), which were found in the dystrophic muscle.[46-49]

The results obtained from dystrophic muscle provide an additional example of the striking correlation between the tissue level of Glc-1,6-P_2 and the activities of the key enzymes of glucose metabolism which are affected by this modulator. The decrease in the level of Glc-

1,6-P_2 in the dystrophic muscle, which is a common regulator of phosphoglucomutase, phosphofructokinase, hexokinase, and 6-phosphogluconate dehydrogenase, was associated with correlated changes in the activities of all four enzymes. Through its regulatory action on the activities of the key enzymes in glucose metabolism, this modulator may regulate the different pathways of glucose metabolism.

In the dystrophic muscle there is a marked accumulation of intracellular Ca^{2+},[50] and it has been postulated that the marked rise in free intracellular Ca^{2+} promotes muscle damage. When intracellular Ca^{2+} was raised experimentally in normal muscle by treatment with Ca^{2+}-ionophore A23187, ultrastructural damage and myofilaments degradation followed quickly.[51] As mentioned above, Ca^{2+}-ionophore A23187 exerts a marked decrease in Glc-1,6-P_2 level in normal muscle.[23,25] This finding suggests that the decrease in Glc-1,6-P_2 concentration in the dystrophic muscle may be mediated by the rise in free intracellular Ca^{2+}. We also found in the dystrophic muscle a significant increase in the level of cyclic GMP.[38] Cyclic GMP has a dual action on carbohydrate metabolism: it induces a decrease in Glc-1,6-P_2 level in muscle,[23,24] and it is also a direct allosteric inhibitor of phosphofructokinase.[4] It seems, therefore, that the increase in the level of this inhibitor, as well as the decrease in the level of Glc-1,6-P_2 (one of the stronger activators of phosphofructokinase), may account for the striking reduction in the phosphofructokinase activity,[38] and thus in glycolysis,[52-54] in dystrophic muscle.

The marked reduction in phosphofructokinase activity in muscular dystrophy is reversible. By adding Glc-1,6-P_2 to the reaction mixture of phosphofructokinase from dystrophic muscle, we were able to activate this enzyme significantly,[38] which indicates that there is no defect in the enzyme's protein. These results strongly suggest that the pathological reduction in phosphofructokinase activity and thus in glycolysis in muscular dystrophy can be reversed by raising intracellular Glc-1,6-P_2 concentration, and thereby increasing phosphofructokinase activity and glycolysis. The reduction in the glycolytic capacity of the dystrophic muscle may lead to the reported decrease in the level of ATP[55,56] (as oxidative metabolism remains intact[52,54]). Since it has been reported by several investigators that conditions which interfere with the supply of ATP may lead to abnormal permeability of cell membranes,[57,60] it is possible that the reduced glycolytic capacity contributes to the pathological membrane changes in muscular dystrophy. Our results strongly suggest that the decrease in Glc-1,6-P_2 may account for the pathological changes in glucose metabolism in dystrophic muscle and may prove to be of fundamental importance in the pathogenesis of muscular dystrophy. Our goal is therefore to find a way to raise intracellular Glc-1,6-P_2 concentration in the dystrophic muscle and thereby correct the abnormalities in glucose metabolism.

D. The Effect of Fasting and Refeeding on Glc-1,6-P_2 and on the Activities of Key Enzymes in Normal and Dystrophic Muscle

Fasting induces a decrease in Glc-1,6-P_2 in muscle, and refeeding reverses this reduction.[39,61] Our experiments[39] have revealed that the levels of Glc-1,6-P_2 in muscle of normal mice are markedly reduced after 22 hr of fasting. When the fasted normal mice were refed for 9 hr, the Glc-1,6-P_2 returned to initial levels. In muscle of dystrophic mice, where the Glc-1,6-P_2 levels were much lower than normal, 22 hr fasting induced a proportionally similar reduction in Glc-1,6-P_2. However, when the fasted dystrophic mice were refed, the Glc-1,6-P_2 levels were not only restored to the initial dystrophic levels, but dramatically reached the levels of normal muscle (Figure 3).[39] Like in other conditions a close correlation was found between tissue levels of Glc-1,6-P_2 and the activities of phosphofructokinase and phosphoglucomutase, which are activated by this regulator. As shown in Tables 2 and 3 phosphofructokinase and phosphoglucomutase activities, when assayed under regulatory conditions, where the enzymes are sensitive to activation by Glc-1,6-P_2,[62] changed in full correlation to Glc-1,6-P_2 levels (Figure 3). The activities of these enzymes (which are high

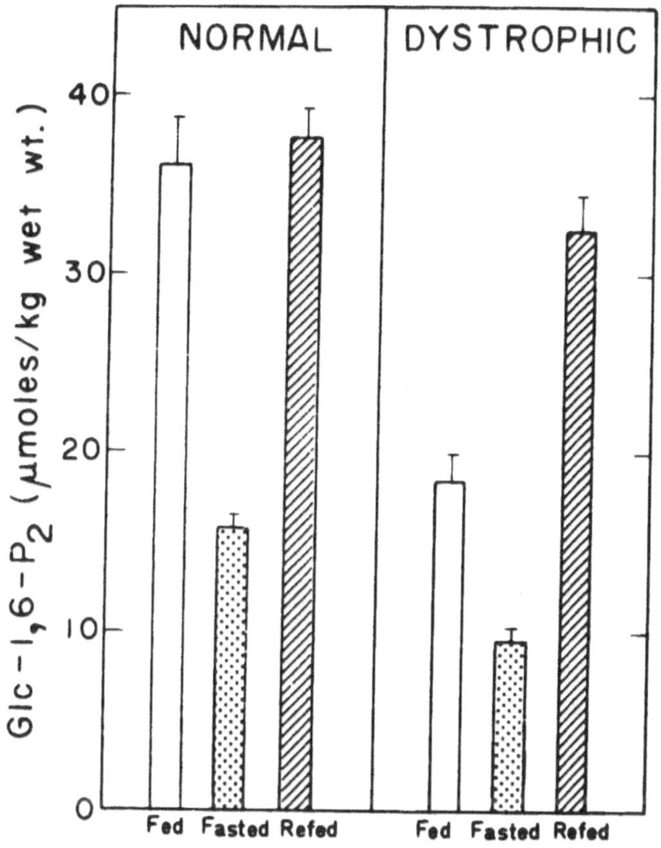

FIGURE 3. The effect of fasting and refeeding on Glc-1,6-P_2 levels in muscle of normal and dystrophic mice. Values are means ± SEM for five to ten animals in each group. *p* Values (fasted vs. fed; refed vs. fasted; dystrophic fed vs. normal fed) <0.005. (From Beitner, R., Nordenberg, J., and Cohen, T. J., *FEBS Lett.*, 104, 244, 1979. With permission.)

in normal and markedly reduced in dystrophy), were reduced by fasting and increased by refeeding, both in the normal and dystrophic mice. However, the dystrophic muscle was much more sensitive to refeeding than the normal muscle. Refeeding of the fasted dystrophic mice induced a dramatic enhancement in the regulatory activities of phosphofructokinase and phosphoglucomutase, reachng values of normal control mice (Tables 2 and 3), in correlation to the overshooting of Glc-1,6-P_2 (Figure 3). These results suggest that refeeding may reverse the reduction in glycolytic capacity of the dystrophic muscle, by raising the concentration of Glc-1,6-P_2.

The fact that the maximal activities of both enzymes (Tables 2 and 3), measured under optimal, nonphysiological conditions, where the enzymes are not subject to regulation by Glc-1,6-P_2 [62] were not changed in dystrophy or by fasting and refeeding, indicates that the total protein content of these enzymes in the cell was not changed through all the different experimental states. The changes in enzymatic activities (assayed under regulatory conditions) may therefore be attributed to the changes in the tissue level of Glc-1,6-P_2.

E. The Effect of Phospholipase A_2 and Lysolecithin

Evidence is now accumulating which points to the possibility of a membrane defect in human and animal muscular dystrophy.[63,64] It has been found that phospholipase A_2 is highly

Table 2
THE EFFECT OF FASTING AND REFEEDING ON PHOSPHOFRUCTOKINASE ACTIVITY IN NORMAL AND DYSTROPHIC MUSCLE

	Phosphofructokinase activity			
	Regulatory activity		Maximal activity	
	mU/mg protein	mU/mg NCP[a]	mU/mg protein	mU/mg NCP[a]
		Normal		
Fed	230 ± 12	283 ± 20	802 ± 23	987 ± 29
Fasted	118 ± 10	153 ± 11	749 ± 14	918 ± 12
Refed	287 ± 24	369 ± 33	740 ± 54	987 ± 69
		Dystrophic		
Fed	63 ± 5	81 ± 6	706 ± 15	918 ± 84
Fasted	27 ± 3	36 ± 3	641 ± 95	812 ± 96
Refed	162 ± 11	223 ± 21	677 ± 36	923 ± 37

Note: Values are means ± SEM for six to ten animals in each group. *p* values: regulatory activity (fasted vs. fed; refed vs. fasted; dystrophic fed vs. normal fed) <0.005; maximal activity, not significant.

[a] NCP, noncollagen protein.

From Beitner, R., Nordenberg, J., and Cohen, T. J., *FEBS Lett.*, 104, 244, 1979. With permission.

(about fivefold) elevated in dystrophic mouse muscle.[65] A decrease in phosphatidylcholine and an increase in lysolecithin in dystrophic muscle has also been reported.[66] All these findings prompted us to examine the possibility that the changes in glucose metabolism which we found in dystrophic muscle may be related to the changes in the activity of phospholipase A_2 and the resulting changes in membrane phospholipids. In order to test this hypothesis we have chosen as a model the normal isolated rat diaphragm muscle. The muscles were incubated in the presence of phospholipase A_2, which has been reported to attack membrane phospholipids under incubation with various cells and tissues (for reviews see Jain,[67] Meldrum[68]). These experiments revealed that phospholipase A_2 induced a marked decrease in the concentration of Glc-1,6-P_2 in the diaphragm muscle.[40] The fall in Glc-1,6-P_2 was accompanied by a concomitant reduction in the activities of phosphofructokinase and phosphoglucomutase, when assayed under conditions in which these enzymes are sensitive to activation by Glc-1,6-P_2. Lysolecithin, the product of of phospholipase A_2 action on membrane phospholipids, mimicked all these effects of phospholipase A_2.[40]

These experiments reveal that phospholipase A_2 or lysolecithin induce changes in normal muscle which mimic those found in the dystrophic muscle. These results suggest that the decrease in Glc-1,6-P_2 and the resulting alterations in carbohydrate metabolism in the dystrophic muscle may be attributed to the marked activation of phospholipase A_2 which occurs in muscular dystrophy,[65] causing abnormalities in membrane phospholipids.

A relationship between phospholipase A_2 activity and glycolysis could be deduced from the early works of Chain,[69,70] who has shown that various snake venoms, of which phospholipase A_2 is an important constituent, inhibit glycolysis in muscle. In this regard it is

Table 3
THE EFFECT OF FASTING AND REFEEDING ON THE ACTIVITY OF PHOSPHOGLUCOMUTASE IN NORMAL AND DYSTROPHIC MUSCLE

	Phosphoglucomutase activity			
	Regulatory activity		Maximal activity	
	mU/mg protein	mU/mg NCP[a]	mU/mg protein	mU/mg NCP[a]
		Normal		
Fed	256 ± 13	315 ± 20	2350 ± 81	2830 ± 130
Fasted	87 ± 9	129 ± 16	2350 ± 100	2870 ± 100
Refed	280 ± 14	371 ± 17	2300 ± 70	2870 ± 150
		Dystrophic		
Fed	132 ± 20	169 ± 30	2266 ± 348	2750 ± 443
Fasted	88 ± 11	111 ± 16	2178 ± 228	2650 ± 320
Refed	177 ± 20	227 ± 27	2157 ± 90	2600 ± 112

Note: Values are mean ± SEM for six to ten animals in each group, p values: regulatory activity (fasted vs. fed; refed vs. fasted; dystrophic fed vs. normal fed) <0.002; maximal activity, not significant.

[a] NCP, noncollagen protein.

From Beitner, R., Nordenberg, J., and Cohen, T. J., *FEBS Lett.*, 104, 244, 1979. With permission.

also interesting that victims surviving sea snake poisoning were reported to regain muscle power very slowly and may exhibit dystrophic signs for several months.[68,71,72]

Our results suggest that a close relationship exists between phospholipase A_2 action on membrane phospholipids and the regulation of carbohydrate metabolism in muscle mediated by Glc-1,6-P_2.

F. Effect of Anoxia and Aerobiosis

Anoxia induces a rapid fall in the intracellular concentration of Glc-1,6-P_2. This effect of anoxia was observed by us in the isolated rat diaphragm muscle,[73] as well as by Passonneau et al.[74] in mouse brain following decapitation.

The effect of anoxia on the levels of Glc-1,6-P_2 in the rat diaphragm are shown in Figure 4. In these experiments paired hemidiaphragms were incubated under aerobic (O_2) conditions, in which the gas phase was 95% O_2/5% CO_2, and compared to those incubated under anaerobic conditions (N_2), where the gas phase was 95% N_2/5% CO_2. These experiments reveal that anoxia induced a significant, time-dependent decrease in Glc-1,6-P_2 concentration in the diaphragm muscle. The fall in the concentration of Glc-1,6-P_2, the potent inhibitor of hexokinase, was accompanied by an activation of this enzyme.[73] The decrease in Glc-1,6-P_2 and the concomitant activation of hexokinase induced by anoxia could be promptly and almost completely reversed by restoration of aerobic conditions. In these experiments we also measured the changes in the concentration of glucose 6-phosphate, the second known inhibitor of hexokinase. Most strikingly, the levels of this regulator were not reduced but rather elevated during the initial 10 min of anoxic incubation. It is thus clear that the activation of hexokinase after 10 min of anoxia could not be attributed to glucose 6-phosphate,

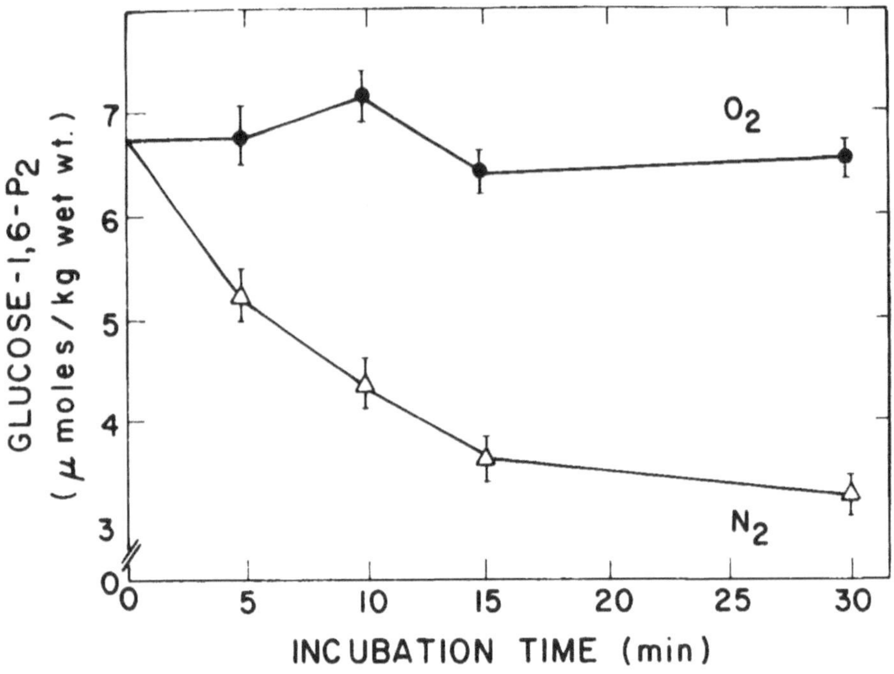

FIGURE 4. Effect of anoxia on Glc-1,6-P_2 levels in the isolated rat diaphragm. For each experiment paired hemidiaphragms pooled from two rats were incubated under aerobic (O_2) or anaerobic (N_2) conditions for 5 to 30 min. Vertical bars represent standard errors of the mean for six to eight experiments. p Values (anaerobic vs. aerobic, at each time of incubation) <0.002. (From Beitner, R., Cohen, T. J., Nordenberg, J., and Haberman, S., *Biochim. Biophys. Acta*, 586, 266, 1979. With permission.)

but was most probably mediated by the fall in Glc-1,6-P_2. Hexokinase was shown to be sensitive to inhibition by Glc-1,6-P_2 at the range of concentration changes during anoxic-aerobic transition.[73] The inverse changes in tissue level of Glc-1,6-P_2 and hexokinase activity during anoxia and aerobiosis, suggest that Glc-1,6-P_2 is the chief regulator which controls the activity of hexokinase to suit the immediate demands of the cell during the aerobic-anoxic transition.

We have also shown that the activity of diaphragm phosphofructokinase is increased during anoxia and inhibited by restoration of aerobiosis.[73] These changes in the enzyme's activity are mediated by changes in the concentrations of its various allosteric effectors, namely, ATP, ADP, AMP, P_i, fructose 1,6-bisphosphate, and fructose 6-phosphate[73] and probably also others. It has been reported that the effectors act in a synergistic manner.[75-78] Thus with anoxia, the reduction in Glc-1,6-P_2, which is also a positive effector of phosphofructokinase, would have little influence on this enzyme since it is activated by the combined action of its other numerous effectors.

G. The Control of Glc-1,6-P_2 by Developmental State and During Growth and Aging

1. Glc-1,6-P_2 During Differentiation and Growth

In embryonic muscle cells in culture the concentration of Glc-1,6-P_2 was found to increase after fusion.[23] This may reflect the differentiation of embryonic muscle tissue from mononucleate myoblasts to multinucleate myotubes (the precursors of muscle fibers) and the accompanying changes in energy metabolism.[23,79]

In the whole animal, we have found that the concentration of Glc-1,6-P_2 in skeletal muscle increases markedly with growth of the rat (Figure 5).[62] Thus in adult rats, at 14 to 16 weeks

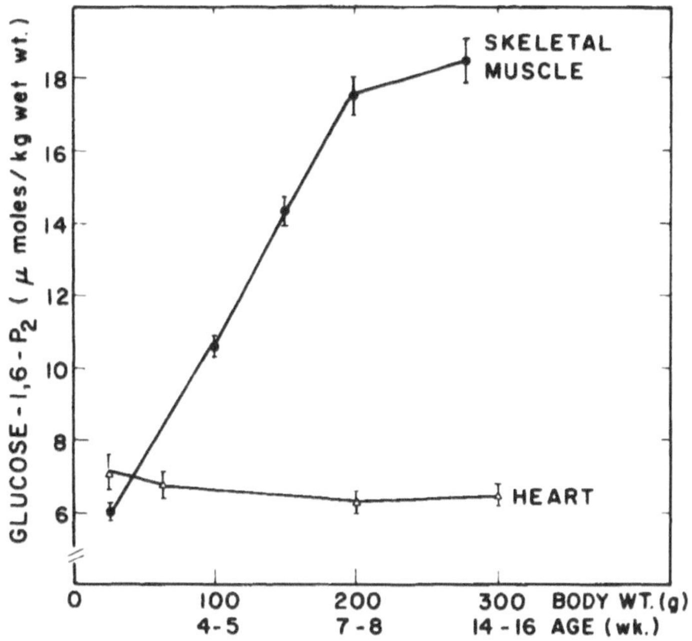

FIGURE 5. Effect of age on Glc-1,6-P_2 levels in rat skeletal muscle and heart. Vertical bars represent standard errors of the mean for 6 to 11 rats per group. p Values (each age point vs. 25 g): skeletal muscle: $p < 0.0005$; heart: not significant. (From Beitner, R., Nordenberg, J., and Cohen, T. J., *Int. J. Biochem.*, 10, 603, 1979. With permission.)

of age (300-g body weight), the skeletal muscle Glc-1,6-P_2 levels were about three times higher than in young animals aged 2 weeks (25-g body weight). In contrast to skeletal muscle, the rat heart exhibited no age-related changes in the level of Glc-1,6-P_2 (Figure 5).

Like in other conditions, a striking correlation was found between the changes in the tissue concentration of Glc-1,6-P_2 and the activities of the key enzymes which are modulated by this regulator. As shown in Figures 6 and 7, in skeletal muscle, concomitant to the age-dependent increase in Glc-1,6-P_2, the potent activator of phosphofructokinase and phosphoglucomutase, the activities of these enzymes increased markedly with growth of the rat. On the other hand in heart, in which the Glc-1,6-P_2 levels remained unchanged during growth (Figure 5), there was also no change in the activity of either phosphofructokinase or phosphoglucomutase. Similarly, the activity of hexokinase correlated with the changes in Glc-1,6-P_2, the potent inhibitor of this enzyme. As shown in Figure 8, hexokinase activity was reduced in skeletal muscle as a function of age, concomitant to the increase in muscle Glc-1,6-P_2 and remained unchanged in heart where the levels of Glc-1,6-P_2 were not changed with age. Thus the activities of all three enzymes which are regulated by Glc-1,6-P_2 correlated with the age-related changes in the levels of this regulator.

The striking correlation between the age-dependent changes in Glc-1,6-P_2 and the activities of phosphofructokinase, phosphoglucomutase, and hexokinase, strongly suggest that the changes in the enzyme activities are mediated by changes in the tissue levels of Glc-1,6-P_2.

In brain, on the other hand, we found a marked decrease in Glc-1,6-P_2 level with growth of the rat.[80] The concentration of Glc-1,6-P_2 in brains of adult rats (5 months of age) was 60% lower as compared to young rats (10 to 14 days of age). This age-related decrease in Glc-1,6-P_2 was accompanied by a correlated increase in the activity of hexokinase and a

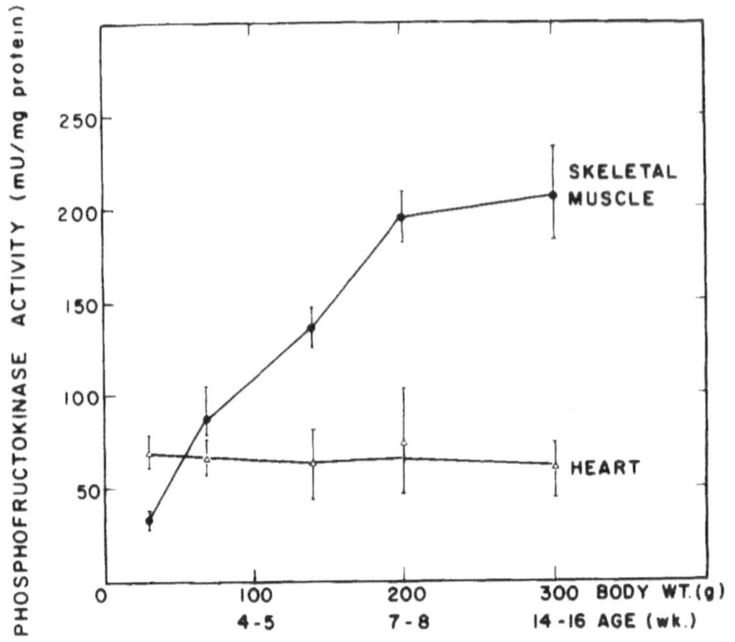

FIGURE 6. The influence of age on phosphofructokinase activity in rat skeletal muscle and heart. Activity was assayed under regulatory conditions (pH 6.9). Values are means ± SE for six to ten rats per group. p Values (each age point vs. 25 g): skeletal muscle: $p < .005$; heart: not significant. (From Beitner, R., Nordenberg, J., and Cohen, T. J., *Int. J. Biochem.*, 10, 603, 1979. With permission.)

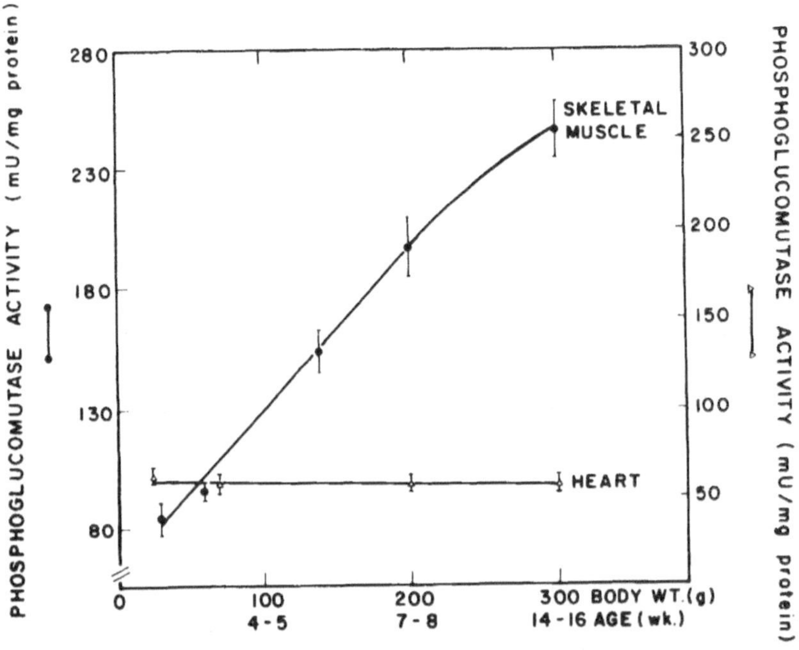

FIGURE 7. The effect of age on phosphoglucomutase activity in rat skeletal muscle and heart. Values are means ± SE for 6 to 13 rats per group. p Values (each age point vs. 25 g); skeletal muscle: $p < 0.005$; heart: not significant. (From Beitner, R., Nordenberg, J., and Cohen, T. J., *Int. J. Biochem.*, 10, 603, 1979. With permission.)

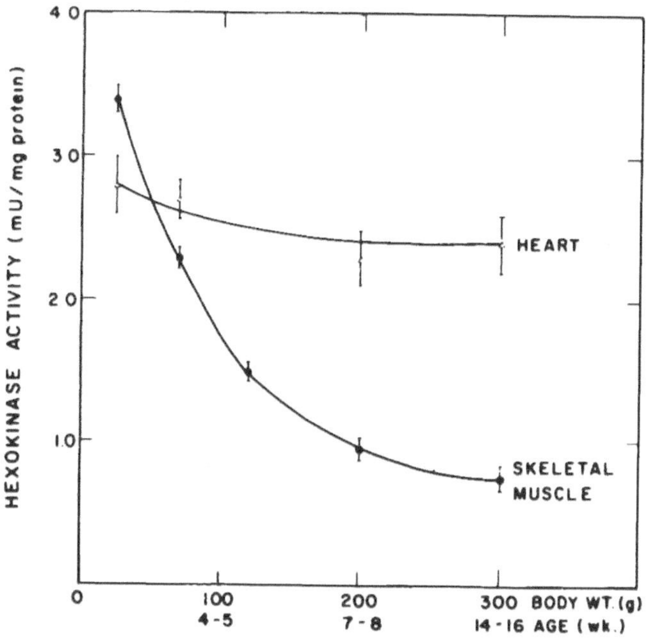

FIGURE 8. The influence of age on hexokinase activity in rat skeletal muscle and heart. Values are means ± SE for five to nine rats per group. p Values (each age point vs. 25 g): skeletal muscle: $p < 0.001$; heart: not significant. (From Beitner, R., Nordenberg, J., and Cohen, T. J., *Int. J. Biochem.*, 10, 603, 1979. With permission.)

reduction in phosphoglucomutase.[80] On the other hand, phosphofructokinase was reported to increase in brain during maturation. However, this increase was attributed to an elevation in the synthesis of the enzyme.[81]

Hexokinase was found to be sensitive to inhibition by Glc-1,6-P_2 at the physiological concentration range found in brain.[80] The enzyme from brains of adult rats was more sensitive to inhibition by low concentrations of Glc-1,6-P_2 as compared to young, most probably as a result of the lower tissue level of Glc-1,6-P_2. The addition of increasing concentrations of Glc-1,6-P_2 to the assay mixture of hexokinase gradually diminished the difference between the adult and young hexokinase activity.[80]

The age-related changes in Glc-1,6-P_2 and in the enzymes' activities in brain were opposite to those found in skeletal muscle. These results suggest that Glc-1,6-P_2 is involved in the regulation of carbohydrate metabolism during growth in both brain and skeletal muscle, and it also regulates the interrelationship between these tissues. The changes in the enzymes' activities in muscle indicate that with growth the ability of the skeletal muscle to utilize glucose is reduced, because hexokinase activity is markedly decreased, whereas its glycogenolytic and glycolytic ability are increased, due to the marked activation of phosphoglucomutase and phosphofructokinase. Namely, the muscle becomes more dependent on glycogen stores as an energy source and uses less glucose from blood, saving it for the brain. In brain, the increase in hexokinase activity, which is partially mediated by the fall in Glc-1,6-P_2, enables a more efficient utilization of blood glucose as an energy source.

2. *Glc-1,6-P_2 in Old Age*

The levels of Glc-1,6-P_2, which increase in skeletal muscle during growth, were found to be strikingly decreased in muscle of old rats (3 years of age).[82] In contrast to the situation during growth, in old age, the fall in Glc-1,6-P_2 affects only phosphoglucomutase activity, causing a severe reduction in this enzyme's activity.

In muscle of old rats, despite the fall in Glc-1,6-P_2, the potent inhibitor of hexokinase, the activity of this enzyme is not increased, although it retains its sensitivity to regulation by Glc-1,6-P_2. This may result from the decrease in the total protein content of this enzyme as reflected in the decrease of its maximal activity.[82]

In contrast to the reduction in hexokinase and phosphoglucomutase in senescence, the regulatory activity of phosphofructokinase, the rate-limiting enzyme in glycolysis, is not reduced in old age, but rather elevated as compared to the adult muscle.[82] This increase results most probably from the increase in the activators AMP and ADP[83,84] and the decrease in its inhibitor, ATP. The activation of this enzyme seems to be a compensatory mechanism which enables the old animal to utilize efficiently the small amount of glucose undergoing phosphorylation and the small amounts of glycogen stores, to provide energy via the glycolytic pathway. Glycolysis becomes a major source of energy in aging muscle as old animals were reported to have a limited aerobic capacity of ATP production.[84] The increase in the activity of phosphofructokinase in muscle of old rats may explain the reported increase in lactate.[84] It is also compatible with the findings that a shift from red (aerobic) towards a more white (anaerobic)-fiber type population occurs during aging.[83,84] Oxygen consumption is reduced in the old animals.[84] It is possible that the lack of oxygen may account for the changes in adenine nucleotides levels, resulting in activation of phosphofructokinase. We also found a marked rise in plasma creatine phosphokinase in old animals,[82] which may point to membrane abnormalities.

H. Effects of Pharmacologic Agents on Cellular Glc-1,6-P_2 Levels

1. The Effects of Local Anesthetics

Local anesthetics, lidocaine and procaine, were found to induce a marked decrease in the level of Glc-1,6-P_2 in the isolated rat diaphragm.[41] The decrease in Glc-1,6-P_2 was accompanied by a concomitant reduction in the activities of phosphofructokinase and phosphoglucomutase. Both local anesthetics also exerted a decrease in the concentration of ATP.[41] All these changes, induced by the local anesthetics, closely resemble the alterations in carbohydrate metabolism which we found in the dystrophic muscle,[38,39,45] or in the normal diaphragm muscle treated with phospholipase A_2, lysolecithin,[40] or Ca^{2+}-ionophore A23187.[25] All these conditions are associated with ultrastructural damage of muscle. It was shown that local anesthetics produced rapid destruction of striated muscle fibers.[86-88]

In the dystrophic muscle, heavy accumulation of Ca^{2+} occurs,[50] and it has been postulated that the marked rise in free intracellular Ca^{2+} promotes muscle damage.[50] Our experiments have revealed that Ca^{2+}-ionophore A23187, which increases intracellular concentration of Ca^{2+} (Reference 89) and causes muscle damage,[51,90] produces the same changes in glucose metabolism and ATP level in the normal diaphragm muscle as those occurring in dystrophic muscle or those induced by the local anesthetics.[25] As local anesthetics were reported to cause accumulation of free Ca^{2+} within the muscle cell,[86,87] it is possible that Ca^{2+} mediates the effects of local anesthetics on Glc-1,6-P_2 and carbohydrate metabolism.

2. Effects of Lithium

Lithium, a drug used widely in the therapy of manic depression, was found to induce a sharp decrease in Glc-1,6-P_2 levels in brain[91,92] and muscles.[92] Following a single Li^+ injection the concentration of Glc-1,6-P_2 in brain, as well as in the tibialis anterior muscle and gastrocnemius muscle of rats, was strikingly reduced.[92] In contrast to its effects in muscles and brain, Li^+ did not affect Glc-1,6-P_2 concentrations in liver of these rats. The decrease in Glc-1,6-P_2 level in brain and muscles induced by Li^+ was accompanied by a marked decrease in the regulatory activity of phosphofructokinase, with no change in its maximal activity. The regulatory activity was measured at pH 6.9 under conditions in which the enzyme was subject to regulation by Glc-1,6-P_2 and its other allosteric modulators. Maximal activity was measured at pH 8.2, under optimal conditions, where the enzyme shows no regulatory (allosteric) properties. The maximal activity reflects the full catalytic potential of the enzyme and the fact that it was not affected by Li^+ indicates that there was no decrease in the total protein content of the enzyme in the cell. In liver, where Glc-1,6-P_2 level was not affected by Li^+, there was also no effect of Li^+ on the activity of phosphofructokinase.[92]

Phosphoglucomutase from muscles and brain showed a similar behavior to that of phosphofructokinase.[92]

The effect of Li^+ was also demonstrated in vitro in the isolated rat diaphragm. Incubation of rat diaphragm muscle in the presence of Li^+, resulted in a sharp decrease in the level of Glc-1,6-P_2.[92] Similar to the findings in vivo, the decrease in Glc-1,6-P_2, the most potent activator of phosphofructokinase and phosphoglucomutase, was accompanied by a marked reduction in the activities of both enzymes, when assayed in the absence of exogenous Glc-1,6-P_2 under conditions in which these enzymes are sensitive to regulation by endogenous Glc-1,6-P_2.

These results show that lithium exerts a marked decrease in Glc-1,6-P_2 concentration in brain and muscles leading to inhibition of phosphofructokinase, the rate-limiting enzyme in glycolysis. It is possible that the reduction in phosphofructokinase and thus in glycolysis induced by lithium, may play an important role in the mechanism of the therapeutic action of this agent in the manic state.

In the dystrophic muscle, where Glc-1,6-P_2 is already markedly reduced and phosphofructokinase is severely depressed,[38] lithium caused an additional decrease in the level of this regulator.[93] It seems, therefore, that lithium treatment of patients afflicted with muscular dystrophy, may result in a more severe depression of the already depressed glycolysis in the dystrophic muscle, and thus may aggravate the muscle damage in this disease.

3. Effects of Trifluoperazine

Recent experiments in our laboratory[94] have shown that injection of trifluoperazine to rats, in therapeutic doses used in treatments of psychotic conditions, induced a significant rise in the level of Glc-1,6-P_2 in muscle. This increase in Glc-1,6-P_2 was accompanied by a concomitant activation of phosphofructokinase and phosphoglucomutase when assayed under conditions in which these enzymes are sensitive to regulation by Glc-1,6-P_2.

I. Glucose-Induced Accumulation of Glc-1,6-P_2 in Pancreatic Islets

Sener et al.[95] have recently reported that a rise in extracellular concentration of glucose caused an increase in the Glc-1,6-P_2 content in rat pancreatic islets. They also found that Glc-1,6-P_2 activated phosphofructokinase in islet homogenates. Based on these findings they have postulated that Glc-1,6-P_2 participates, together with fructose 2,6-bisphosphate, in the activation of phosphofructokinase in intact islet cells, with a resulting increase in both glycolytic flux and insulin secretory rate.[95]

IV. BIOSYNTHESIS AND DEGRADATION OF GLC-1,6-P_2

A. Biosynthesis

A specific enzyme which synthesizes Glc-1,6-P_2, the Glc-1,6-P_2 synthase, was purified from brain.[96,97] There is, however, no information on the occurrence of this enzyme in muscle. Several mechanisms have been proposed for the biosynthesis of Glc-1,6-P_2 in muscle: through the side reaction of phosphoglucomutase with fructose 1,6-bisphosphate[74] or with 1,3-diphosphoglycerate;[18,98] through the side reaction of phosphofructokinase with glucose

Table 4
FACTORS AND CONDITIONS WHICH AFFECT THE ACTIVITY OF GLUCOSE 1,6-BISPHOSPHATASE IN MUSCLE

Factors or conditions	Changes in glucose 1,6-bisphosphatase activity[a]	Ref.
Dibutyryl cAMP	↓	24
Dibutyryl cGMP	↑	24
Ca^{2+}	↑	106
Calmodulin	↑	106
Trifluoroperazine	↓	94, 106
Mn^{2+}	↑	106
Sr^{2+}	↑	106
Phospholipase A_2	↑	40
Lysolecithin	↑	40
Muscular dystrophy	↑	39, 105
Fasting		
Normal mice	↑	39
Dystrophic mice	↑	39
Refeeding		
Normal mice	↓	39
Dystrophic mice	↓	39
Aging	↑	82
Bradykinin	↑	43

[a] ↑, increase; ↓, decrease.

1-phosphate and ATP,[99,100] known previously as "glucose-1-P kinase".[101] A glucose-1-phosphate transphosphorylase reaction has also been reported.[102]

B. Degradation

A specific enzyme which degrades Glc-1,6-P_2, the glucose-1,6-bisphosphatase, was demonstrated in liver,[103,104] muscle,[24,39,40,82,105,106] brain,[107] and skin.[116]

We have studied the activity of this enzyme in muscle, using a sensitive fluorometric method and found that the enzyme is active at physiological pH, with no activity at acid pH.[105]

As summarized in Table 4 the activity of glucose-1,6-bisphosphatase in muscle is controlled by many factors and conditions and may account for the changes in the intracellular concentration of Glc-1,6-P_2 under these conditions.

1. The Control of Glucose-1,6-Bisphosphatase by Cyclic Nucleotides

It was shown that the addition of dibutyryl cyclic AMP to the incubation medium of rat diaphragm muscles caused a significant reduction in the activity of glucose-1,6-bisphosphatase.[24] This reduction may account for the rise in Glc-1,6-P_2 levels induced by dibutyryl cyclic AMP under identical conditions.[13] The effect of dibutyryl cyclic AMP was most pronounced in hemidiaphragms from starved rats, in which the activity of glucose-1,6-bisphosphatase is markedly elevated.

On the other hand dibutyryl cyclic GMP induced an effect opposite to that of dibutyryl cyclic AMP when added to the incubation medium of hemidiaphragms from fed rats; it caused a marked increase in the activity of glucose-1,6-bisphosphatase.[24] This increase may account for the decrease in the intracellular concentration of Glc-1,6-P_2 induced by dibutyryl cyclic GMP.[24] Thus, the two cyclic nucleotides act in an "Yin-Yang" manner,[108,109] exerting opposite influences on the activity of glucose-1,6-bisphosphatase and thereby on the levels of Glc-1,6-P_2.

2. The Control of Glucose-1,6-Bisphosphatase by Ca^{2+} and Calmodulin

It has been recently found by Wakelam et al.[106] that Ca^{2+} and calmodulin control the activity of glucose-1,6-bisphosphatase. The enzyme from rat skeletal muscle extract was found to be activated by physiological Ca^{2+} concentrations that may be achieved in a cell after stimulation by hormones such as vasopressin, or adrenaline acting through α-receptors. Ca^{2+} could be replaced by Sr^{2+} and Mn^{2+}, but their effects were less pronounced. The activation of the enzyme by Ca^{2+} was inhibited by EGTA. The addition of trifluoperazine, an inhibitor of calmodulin function, also inhibited the enzyme's activity, this effect could be reversed by adding exogenous calmodulin.[106] Recent experiments in our laboratory[94] have shown that trifluoperazine could also act in vivo. Injection of this antidepressive phenothiazine derivative to rats induced inhibition of glucose-1,6-bisphosphatase in vivo, which is compatible with the in vitro experiments of Pette and co-workers.[106] Inhibition of glucose-1,6-bisphosphatase following the injection of trifluoperazine may account for the increase in Glc-1,6-P_2 level induced by this compound.[94]

The activation of glucose-1,6-bisphosphatase by Ca^{2+} and calmodulin may be the mechanism through which hormones, such as vasopressin, serotonin, bradykinin, or adrenaline acting through α-receptors, decrease the muscle Glc-1,6-P_2 level. The decrease of Glc-1,6-P_2 concentration in dystrophic muscle or in normal muscle treated with Ca^{2+}-ionophore A23187, or local anesthetics (which increase free intracellular Ca^{2+}), may also be mediated through activation of glucose-1,6-bisphosphatase by Ca^{2+} and calmodulin.

3. The Influence of Phospholipase A_2 and Lysolecithin on the Activity of Glucose-1,6-Bisphosphatase

Phospholipase A_2 was found to induce a marked, time-dependent activation of glucose-1,6-bisphosphatase in the isolated rat diaphragm muscle.[40] The activation of glucose-1,6-bisphosphatase may account for the reduction in the levels of Glc-1,6-P_2 exerted by phospholipase A_2 (Figure 9).[40] Lysolecithin, the product of phospholipase A_2 action on membrane phospholipids, mimicked the effects of phospholipase A_2; 30 min incubation in the presence of lysolecithin resulted in an enhancement of the activity of glucose-1,6-bisphosphatase and a decrease in Glc-1,6-P_2 concentration in the diaphragm muscle.[40]

4. Changes in Glucose-1,6-Bisphosphatase Activity in Muscular Dystrophy

In muscle from genetically dystrophic mice we found a marked increase in the activity of glucose-1,6-bisphosphatase.[39,105] The enhancement of this enzyme's activity in the dystrophic mice was observed already at the age of 1 month, when the first signs of muscle weakness were detectable.[105] A further increase in the enzyme's activity was found at the age of 2 months. The activation of glucose-1,6-bisphosphatase could account for the reduction in Glc-1,6-P_2 levels, leading to the pathological changes in glucose metabolism in dystrophic muscle.[38,39,45]

In the dystrophic muscle phospholipase A_2 is highly elevated.[65] There is also a marked rise in free intracellular Ca^{2+},[50] and a marked increase in cyclic GMP level.[38] As shown in Figure 10 each of these factors may lead to activation of glucose-1,6-bisphosphatase.[24,40,106] They may also be interrelated (Figure 10): phospholipase A_2,[110,111] glucose-1,6-bisphosphatase,[106] and guanylate cyclase,[111] are activated by Ca^{2+} and calmodulin. Thus the heavy accumulation of Ca^{2+} in the dystrophic muscle may account for activation of all these three enzymes. An increase in the activity of guanylate cyclase has been reported in muscle of dystrophic mice,[112] which may account for the increase which we found in cyclic GMP concentration.[38] On the other hand, the high activity of phospholipase A_2 in muscular

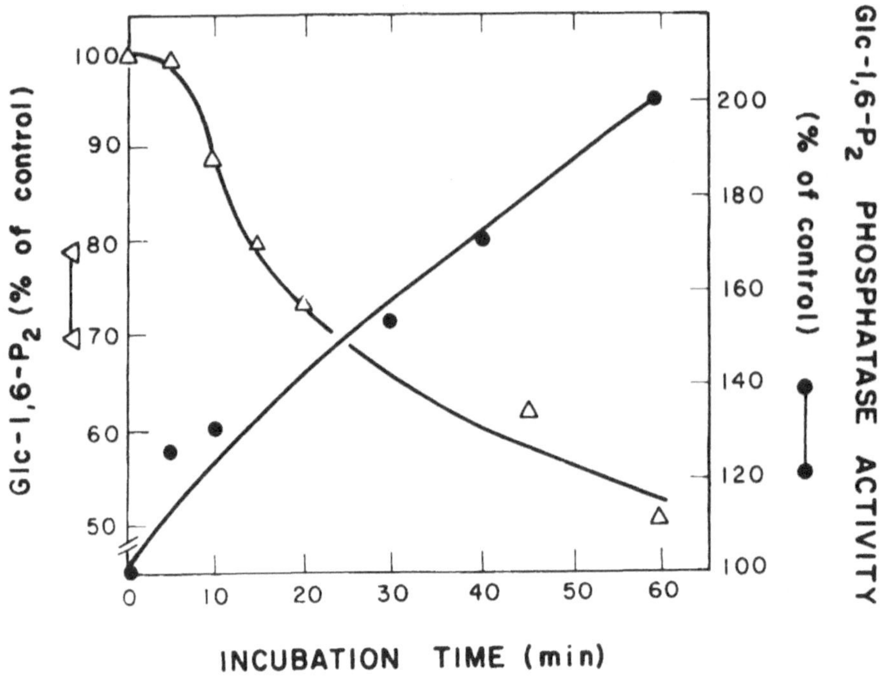

FIGURE 9. The effect of phospholipase A_2 on the levels of Glc-1,6-P_2 (△) and on the activity of glucose 1,6-bisphosphatase (●) in the isolated rat diaphragm. For each experiment paired hemidiaphragms were incubated in the absence and presence of phospholipase A_2 (32 units/mℓ) for 5 to 60 min. (From Beitner, R., Nordenberg, J., Cohen, T. J., and Beery, E., *Int. J. Biochem.*, 11, 467, 1980. With permission.)

dystrophy may cause, through its effect on membrane phospholipids, an enhancement in Ca^{2+} transport through the membrane, leading to an increase in free intracellular Ca^{2+}. Phospholipase A_2 and its product, lysolecithin, were also found to activate guanylate cyclase,[113-115] and glucose-1,6-bisphosphatase.[40]

5. Effect of Fasting and Refeeding on Glucose-1,6-Bisphosphatase Activity in Normal and Dystrophic Muscle

Our experiments revealed that fasting induced an elevation in glucose-1,6-bisphosphatase activity in both normal and dystrophic muscle and refeeding reduced its activity.[39] However, the dystrophic muscle was much more sensitive to refeeding than the normal muscle. The activity of glucose-1,6-bisphosphatase, which rises markedly in the dystrophic muscle, was dramatically reduced to normal levels in the refed dystrophic mice.[39] This offers an explanation to the mechanism by which Glc-1,6-P_2 rises so strikingly following refeeding (Figure 3). The factor responsible for inhibition of this enzyme's activity during refeeding is as yet unknown.

6. Influence of Age on the Activity of Glucose-1,6-Bisphosphatase

The activity of glucose-1,6-bisphosphatase was found to be significantly enhanced in skeletal muscle of old rats (3 years of age).[82] This increase may account for the reduction in the concentration of Glc-1,6-P_2 in the aged animals.[82] The activity of glucose-1,6-bisphosphatase in gastrocnemius muscle was not changed during growth, which suggests that the marked rise in muscle Glc-1,6-P_2 with growth,[62] results from an increased synthesis of this regulator.

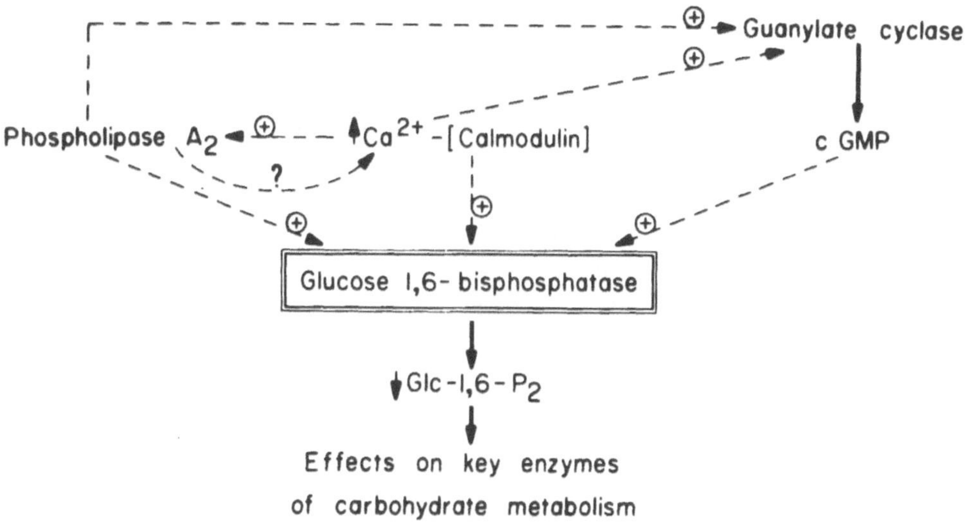

FIGURE 10. Proposed mechanisms for the activation of glucose 1,6-bisphosphatase in muscular dystrophy.

In rat skin, on the other hand, we found a full correlation between changes in Glc-1,6-P_2 level and the activity of glucose-1,6-bisphosphatase during growth.[116] The enzyme's activity in skin was extremely low at 8 to 12 days of age, when the levels of Glc-1,6-P_2 are high, and was markedly enhanced (about fourfold), in skin of adult rats (2 months of age), when the levels of Glc-1,6-P_2 are significantly decreased.[116]

7. Effects of Bradykinin on Glucose-1,6-Bisphosphatase Activity

Injection of bradykinin to rats induced a marked enhancement of glucose-1,6-bisphosphatase activity in skin[42] and in the tibialis anterior muscle.[43] The increase in the activity of this enzyme may account for the reduction in the levels of Glc-1,6-P_2 induced by bradykinin.

The mechanism for the activation of glucose-1,6-bisphosphatase by bradykinin may be similar to that occurring in muscular dystrophy (Figure 10). It has been reported that bradykinin markedly enhances phospholipase A_2 activity.[117] It also increases cyclic GMP concentration[43,44] in a Ca^{2+}-dependent manner.[44] It is thus possible that phospholipase A_2 as well as cyclic GMP and Ca^{2+} may be involved in the bradykinin-induced activation of glucose-1,6-bisphosphatase, as all these three factors were found to activate this enzyme,[24,40,106] and they may also be interrelated as described in Section IV.B.4.

V. CONCLUDING REMARKS

Glc-1,6-P_2 is a powerful regulator which modulates the activities of key enzymes of carbohydrate metabolism (Figure 1). Glc-1,6-P_2 is not a biosynthetic precursor or an intermediate in a metabolic pathway, but a regulatory molecule which acts as a signal and controls the different pathways of carbohydrate metabolism through its regulatory action on the key enzymes in these pathways.

A long list of factors and conditions affect the intracellular levels of Glc-1,6-P_2 (Table 1), resulting in concomitant changes in the activities of the enzymes which are regulated by this modulator. The "second messenger triumvirate", namely cyclic AMP, cyclic GMP, and Ca^{2+} control the cellular concentration of Glc-1,6-P_2: cyclic AMP increases, whereas cyclic GMP and Ca^{2+} decrease the Glc-1,6-P_2 levels. These effects were demonstrated by us in the isolated rat diaphragm muscle,[13,24,25] as well as by Pette and co-workers,[22,23] in cultured muscle tissue. Hormones which act through cyclic AMP, e.g., epinephrine, exert an increase in cellular Glc-1,6-P_2 level[13,23] whereas hormones or neurotransmitters which act through cyclic GMP and Ca^{2+}, e.g., epinephrine + β-blocker, vasopressin, serotonin, and bradykinin, induce a decrease in Glc-1,6-P_2 level.[23,27,28,42,43]

The intracellular Glc-1,6-P_2 levels are also controlled by fasting and refeeding,[39,61] by anoxia and aerobiosis,[73,74] by developmental state,[23] and during growth[62,80,116] and aging.[82] Glucose was reported to induce accumulation of Glc-1,6-P_2 in pancreatic islets.[95]

Various pharmacologic agents, such as lithium,[91-93] local anesthetics,[41] and trifluoperazine[94] also affect cellular Glc-1,6-P_2 levels. The concentration of Glc-1,6-P_2 is severely decreased in muscular dystrophy,[38,39,45,61] or in the normal muscle following treatment with phospholipase A_2 or lysolecithin.[40] Under all these conditions, the changes in cellular Glc-1,6-P_2 levels led to concomitant changes in the activities of the key enzymes from carbohydrate metabolism which are modulated by this regulator.

The changes in Glc-1,6-P_2 levels under most of these conditions could be attributed to changes in the activity of glucose-1,6-bisphosphatase, the enzyme that degrades Glc-1,6-P_2, (Table 4). This enzyme is activated by Ca^{2+} and calmodulin.[106] It is also regulated by cyclic AMP and cyclic GMP,[24] in an "Yin-Yang" manner.[108,109] Phospholipase A_2 was found to activate potently glucose-1,6-bisphosphatase (Figure 9) leading to a decrease in Glc-1,6-P_2 level. The regulation of glucose 1,6-bisphosphatase by Ca^{2+}-calmodulin, cyclic nucleotides, and phospholipase A_2 may be the mechanisms that control cellular Glc-1,6-P_2 levels under different hormonal, physiological, and pathological conditions.

We have shown that in muscular dystrophy, in which phospholipase A_2 activity, intracellular Ca^{2+}, and cyclic GMP increase, the activity of glucose-1,6-bisphosphatase is markedly enhanced.[39,105] This may account for the decrease in Glc-1,6-P_2 leading to the pathological changes in carbohydrate metabolism (Figure 10).[38,39,45]

Changes in Glc-1,6-P_2 and carbohydrate metabolism should also be investigated in other pathological conditions which are characterized by changes in the levels of cyclic nucleotides, calcium, or phospholipase A_2 activity.

Fructose 2,6-bisphosphate, another recently discovered hexose bisphosphate (for reviews see References 118 to 120 and Chapter 2 by Hue and Bartrons in this book), is also a potent activator of phosphofructokinase and an inhibitor of fructose 1,6-bisphosphatase. However, we found no effect of this compound on the activities of hexokinase or 6-phosphogluconate dehydrogenase. It has also no effect on phosphoglucomutase activity.[95] The unique feature of Glc-1,6-P_2 is that it acts as a powerful regulator common to several key enzymes in the different pathways of carbohydrate metabolism (Figure 1), whereas the action of fructose 2,6-bisphosphate is limited to phosphofructokinase and fructose 1,6-bisphosphatase. Fructose 2,6-bisphosphate undoubtedly plays an important regulatory role in liver; however, it has been suggested that this compound may not be an important effector of phosphofructokinase in skeletal muscle.[121] Glc-1,6-P_2 on the other hand is an extremely important regulator of carbohydrate metabolism in muscle, as well as in other tissues. In pancreatic islets both fructose 2,6-bisphosphate and Glc-1,6-P_2 were suggested to play a regulatory role in the activation of phosphofructokinase, acting in a synergistic manner.[95] It is possible that the two regulators may also act synergistically in other tissues and conditions. This requires further investigation.

We have recently found in rat skin a close correlation between age-dependent changes in Glc-1,6-P_2 and the activities of particulate (mitochondrial) hexokinase and 6-phosphogluconate dehydrogenase.[116] The binding of hexokinase to mitochondria has been extensively studied (see Chapters 3 and 4 by Wilson and Rijksen and Staal in this book, and Bessman[122]). Changes in the levels of Glc-1,6-P_2 and the particulate and soluble enzyme under different conditions is at present under investigation in our laboratory.

Our studies on the regulatory action of Glc-1,6-P_2 concentrated chiefly on muscle. The role of Glc-1,6-P_2 should be investigated in other tissues under different physiological and pathological conditions.

The regulatory role of Glc-1,6-P_2 may not be limited to carbohydrate metabolism, but it may also act on protein and lipid metabolism. All these aspects have not yet been investigated.

ACKNOWLEDGMENTS

This work was supported by grants from the United States-Israel Binational Science Foundation; the Muscular Dystrophy Association of America; the Ministry of Industry and Commerce, Chief Scientist's Office, Jerusalem, Israel, and in part by the Research Committee, Bar-Ilan University. I thank Mrs. Bluma Lederhendler and Mrs. Eloy Kaneti for typing this manuscript.

REFERENCES

1. **Beitner, R.,** The role of glucose-1,6-bisphosphate in the regulation of carbohydrate metabolism in muscle, *Trends Biochem. Sci.*, 4, 228, 1979.

1a. **Beitner, R.,** Control of levels of glucose 1,6-bisphosphate, *Int. J. Biochem.*, 16, 579, 1984.

2. **Hofer, H. W. and Pette, D.,** Wirkungen und Wechselwirkungen von Substraten und Effektoren an der Phosphofructokinase des Kaninchen-Skeletmuskels, *Hoppe Seyler's Z. Physiol. Chem.*, 349, 1378, 1968.

3. **Krzanowski, J. and Matschinsky, F. M.,** Regulation of phosphofructokinase by phosphocreatine and phosphorylated glycolytic intermediates, *Biochem. Biophys. Res. Commun.*, 34, 816, 1969.

4. **Beitner, R., Haberman, S., and Cycowitz, T.,** The effect of cyclic GMP on phosphofructokinase from rat tissues, *Biochim. Biophys. Acta*, 482, 330, 1977.

5. **Rapoport, S.,** Control mechanisms of red cell glycolysis, in *The Human Red Cell In Vitro*, Greenwalt, T. J. and Jamieson, G. A., Eds., Grune & Stratton, New York, 1974, 153.

6. **Rose, I. A. and Warms, J. V. B.,** Glucose- and mannose-1,6-P_2 as activators of phosphofructokinase in red blood cells, *Biochem. Biophys. Res. Commun.*, 59, 1333, 1974.

7. **Koster, J. F., Slee, R. G., Staal, G. E. J., and Van Berkel, Th. J. C.,** The influence of glucose-1,6-diphosphate on the enzymatic activity of pyruvate kinase, *Biochem. Biophys. Acta*, 258, 763, 1972.

8. **Beitner, R., Haberman, S., and Livni, L.,** Complementarity in the regulation of phosphoglucomutase, phosphofructokinase and hexokinase: the role of glucose-1,6-bisphospate, *Biochim. Biophys. Acta*, 397, 355, 1975.

9. **Gerber, G., Preissler, H., Heinrich, R., and Rapoport, S. M.,** Hexokinase of human erythrocytes. Purification, kinetic model and its application to the conditions in the cell, *Eur. J. Biochem.*, 45, 39, 1974.

10. **Rose, I. A. and Warms, J. V. B.,** pH dependence of the α-glucose-1,6-diphosphate inhibition of hexokinase-II, *Arch. Biochem. Biophys.*, 171, 678, 1975.

11. **Rijksen, G. and Staal, G. E. J.,** Regulation of human erythrocyte hexokinase by glucose-1,6-diphosphate and inorganic phosphate, *FEBS Lett.*, 80, 61, 1977.

12. **Leloir, L. F., Trucco, R. E., Cardini, C. E., Paladini, A. C., and Caputto, R.,** The coenzyme of phosphoglucomutase, *Arch. Biochem.*, 19, 339, 1948.

13. **Beitner, R., Haberman, S., and Nordenberg, J.,** The effect of epinephrine and dibutryl cyclic AMP on glucose-1,6-bisphosphate levels and the activities of hexokinase, phosphofructokinase and phosphoglucomutase in the isolated rat diaphragm, *Mol. Cell. Endocrinol.*, 10, 135, 1978.

14. **Marcus, C. J.,** Inhibition of bovine hepatic fructose-1,6-diphosphatase, by substate analogs, *J. Biol. Chem.*, 251, 2963, 1976.

15. **Beitner, R. and Nordenberg, J.,** Inhibition of 6-phosphogluconate dehydrogenase (decarboxylating) by glucose-1,6-bisphosphate, *Biochim. Biophys. Acta*, 583, 266, 1979.

16. **Peck, E. J., Jr. and Ray, W. J., Jr.,** Metal complexes of phosphoglucomutase *in vivo*, Alterations induced by insulin, *J. Biol. Chem.*, 246, 1160, 1971.

17. **Johnson, G. B.,** Enzyme polymorphism and metabolism, *Science*, 184, 28, 1974.

18. **Alpers, J. B.,** Studies on phosphomutases. Effects of 1,3-diphosphoglycerate on phosphoglucomutase, *J. Biol. Chem.*, 243, 1698, 1968.

19. **Tarnowski, W., Kittler, M., and Hilz, H.,** Effect of cortisol on the conversion of hexose phosphates into glycogen, *Biochem. Z.*, 341, 45, 1964.
20. **Kohle, S. J. and Vannucci, R. C.,** Glycogen metabolism in fetal and post-natal rat brain; influence of birth, *J. Neurochem.*, 28, 441, 1977.
21. **Sabularse, D. C. and Anderson, R. L.,** Inorganic pyrophosphate: D-fructose-6-phosphate 1-phosphotransferase in mung beans and its activation by D-fructose 1,6-bisphosphate and D-glucose 1,6-bisphosphate, *Biochem. Biophys. Res. Commun.*, 100, 1423, 1981.
22. **Wakelam, M. J. O., Emmerich, M., and Pette, D.,** Responses of glucose-1,6-bisphosphate in myoblasts and myotubes to α and β-hormonal stimulation, *Biochem. Soc. Trans.*, 9, 464, 1981.
23. **Wakelam, M. J. O. and Pette, D.,** The control of glucose-1,6-bisphosphate by developmental state and hormonal stimulation in cultured muscle tissue, *Biochem. J.*, 204, 765, 1982.
24. **Beitner, R. and Cohen, T. J.,** Opposite effects of dibutyryl cyclic GMP and dibutyryl cyclic AMP on glucose-1,6-diphosphate levels and the activities of glucose-1,6-diphosphate phosphatase and phosphofructokinase in diaphragm muscle, *FEBS Lett.*, 115, 197, 1980.
25. **Beery, E., Klein, S., Nordenberg, J., and Beitner, R.,** The influence of Ca^{2+}-ionophore A23187 on glucose-1,6-diphosphate and ATP levels, and on the activities of phosphofructokinase and phosphoglucomutase, in isolated rat diaphragm muscle, *Biochem. Int.*, 1, 526, 1980.
26. **Berridge, M. J.,** Receptors and calcium signalling, *Trends Pharmacol. Sci.*, 1, 419, 1980.
27. **Beitner, R., Kaplansky, M., and Frucht, H.,** Effect of serotonin on glucose-1,6-diphosphate levels in rat muscle and skin, *IRCS Med. Sci.*, 10, 300, 1982.
28. **Beitner, R., Frucht, H., and Kaplansky, M.,** Changes in the levels of glucose-1,6-diphosphate and cyclic GMP, and in the activities of phosphofructokinase and phosphoglucomutase induced by serotonin in muscle, *Int. J. Biochem.*, 15, 935, 1983.

29a. **Mansour, T. E.,** Phosphofructokinase, in *Current Topics in Cellular Regulation*, Vol. 5, Horecker, B. L. and Stadtman, E. R., Eds., Academic Press, New York, 1972, 1.

29b. **Bloxham, D. P. and Lardy, H. A.,** *The Enzymes*, Vol. 8, Boyer, P. D., Ed., Academic Press, New York, 1973, 239.

30. **Uyeda, K.,** Phosphofructokinase, in *Advances in Enzymology*, Vol. 48, Meister, A., Ed., John Wiley & Sons, New York, 1979, 193.
31. **Patten, B. M., Oliver, K. L., and Engel, W. K.,** Serotonin-induced muscle weakness, *Arch. Neurol.*, 31, 347, 1974.
32. **O'Steen, W. K., Barnard, J. L., and Yates, R. D.,** Morphologic changes in skeletal muscle induced by serotonin treatment, *Exp. Mol. Pathol.*, 7, 145, 1967.
33. **Murphy, D. L., Mendell, J. R., and Engel, W. K.,** Serotonin and platelet function in Duchenne muscular dystrophy, *Arch. Neurol.*, 28, 239, 1973.
34. **Hudecki, M. S., Pollina, C. M., Bhargava, A. K., and Hudecki, R. S.,** Screening of antiserotoninergic drugs with the genetically dystrophic chicken, *Arch. Neurol.*, 37, 545, 1980.
35. **Barnard, E. A. and Barnard, P. J.,** Use of genetically dystrophic animals in chemotherapy trials and application of serotonin antagonists as antidystrophic drugs, *Ann. N.Y. Acad. Sci.*, 317, 374, 1979.
36. **Swash, M., Fox, K. P., and Davidson, A. R.,** Carcinoid myopathy. Serotonin-induced muscle weakness in man, *Arch. Neurol.*, 32, 572, 1975.
37. **Berry, E. M., Maunder, C., and Wilson, M.,** Carcinoid myopathy and treatment with cyproheptadine (Periactin), *Gut*, 15, 34, 1974.
38. **Beitner, R., Haberman, S., Nordenberg, J., and Cohen, T. J.,** The levels of cyclic GMP and glucose-1,6-diphosphate and the activity of phosphofructokinase in muscle from normal and dystrophic mice, *Biochim. Biophys. Acta*, 542, 537, 1978.
39. **Beitner, R., Nordenberg, J., and Cohen, T. J.,** Changes in glucose-1,6-diphosphate and in the activities of phosphofructokinase, phosphoglucomutase and glucose 1,6-diphosphate phosphatase induced by fasting and refeeding in dystrophic muscle, *FEBS Lett.*, 104, 244, 1979.
40. **Beitner, R., Nordenberg, J., Cohen, T. J., and Beery, E.,** The effects of phospholipase A and lysolecithin on glucose-1,6-diphosphate levels and on the activities of glucose-1,6-diphosphate phosphatase, phosphofructokinase and phosphoglucomutase in the isolated rat diaphragm muscle, *Int. J. Biochem.*, 11, 467, 1980.
41. **Nordenberg, J., Klein, S., Beery, E., Kaplansky, M., and Beitner, R.,** Changes in the levels of glucose-1,6-diphosphate and ATP and in the activities of phosphofructokinase and phosphoglucomutase induced by local anesthetics in the isolated rat diaphragm muscle, *Int. J. Biochem.*, 13, 1005, 1981.
42. **Lilling, G., Frucht, H., Ben-Porat, H., and Beitner, R.,** Changes in glucose-1,6-bisphosphate levels and in the activity of glucose-1,6-bisphosphatase induced by bradykinin in rat skin, *IRCS Med. Sci.*, 11, 130, 1983.

43. **Frucht, H., Lilling, G., and Beitner, R.,** Influence of bradykinin on glucose 1,6-bisphosphate and cyclic GMP levels and on the activities of glucose 1,6-bisphosphatase, phosphofructokinase and phosphoglucomutase in muscle, *Int. J. Biochem.*, 16, 397, 1984.
44. **Clyman, R. I., Blacksin, A. S., Sandler, J. A., Manganiello, V. C., and Vaughan, M.,** The role of calcium in regulation of cyclic nucleotide content in human umbilical artery, *J. Biol. Chem.*, 250, 4718, 1975.
45. **Beitner, R. and Nordenberg, J.,** The regulatory role of glucose-1,6-diphosphate in muscle of dystrophic mice, *FEBS Lett.*, 98, 199, 1979.
46. **Srivastava, U., Devi, A., and Sarkar, N.,** Biochemical changes in progressive muscular dystrophy. I. Nucleic acid metabolism in normal and dystrophic rabbit and mouse liver, brain and muscle, *Exp. Cell Res.*, 29, 289, 1963.
47. **Srivastava, U.,** Biochemical changes in progressive muscular dystrophy. VII. Studies on the biosynthesis of protein and RNA in various cellular fractions of the muscle of normal and dystrophic mice, *Can. J. Biochem.*, 46, 35, 1968.
48. **Goldspink, D. F. and Goldspink, G.,** Age-related changes in protein turnover and ribonucleic acid of the diaphragm muscle of normal and dystrophic hamsters, *Biochem. J.*, 162, 191, 1977.
49. **Goldspink, D. F.,** Age-related changes of RNA and DNA in muscles of normal and dystrophic hamsters, *Life Sci.*, 20, 57, 1977.
50. **Engel, W. K.,** Integrative histochemical approach to the defect of Duchenne muscular dystrophy, in *Pathogenesis of Human Muscular Dystrophies*, Rowland, L. P., Ed., Excerpta Medica, Amsterdam, 1977, 277.
51. **Duncan, C. J.,** Role of intracellular calcium in promoting muscle damage: a strategy for controlling dystrophic condition, *Experientia*, 34, 1531, 1978.
52. **Dreyfus, J. C., Schapira, G., Schapira, F., and Remos, J.,** Activites enzymatiques du muscle humain. Recherches sur la biochimie comparee de l'homme normal et myopathique, et du rat, *Clin. Chim. Acta*, 1, 434, 1956.
53. **Vignos, P. J., Jr. and Lefkowitz, M.,** A biochemical study of certain skeletal muscle constituents in human progressive muscular dystrophy, *J. Clin. Invest.*, 38, 873, 1959.
54. **Heyck, M., Laudahn, G., and Lüders, C. J.,** Fermentaktivitatsbestimmungen in der gesunden menschlichen Muskulatur und bei Myopathien, *Klin. Wochenschr.*, 41, 500, 1963.
55. **Vignos, P. J., Jr. and Warner, J. L.,** Glycogen, creatine, and high energy phosphate in human muscle disease, *J. Lab. Clin. Med.*, 62, 579, 1963.
56. **Stengel-Rutkowski, L. and Barthelmai, W.,** Über den Muskel-Energie Stoffwechsel bei Kindern mit progressiver Muskeldystrophie typ Duchenne, *Klin. Wochenschr.*, 51, 957, 1973.
57. **Sweetin, J. C. and Thomson, W. H. S.,** Enzyme efflux and clearance, *Clin. Chim. Acta*, 48, 403, 1973.
58. **Wilkinson, J. H. and Robinson, J. M.,** Effect of ATP on release of intracellular enzymes from damaged cells, *Nature (London)*, 249, 662, 1974.
59. **Thomson, W. H. S., Sweetin, J. C., and Hamilton, I. J. D.,** ATP and muscle enzyme efflux after physical exertion, *Clin. Chim. Acta*, 59, 241, 1975.
60. **Gazitt, Y., Ohad, I., and Loyter, A.,** Changes in phospholipid susceptibility toward phospholipases induced by ATP depletion in avian and amphibian erythrocyte membranes, *Biochim. Biophys. Acta*, 382, 65, 1975.
61. **Beitner, R. and Nordenberg, J.,** Effect of fasting and refeeding on glucose-1,6-diphosphate levels in muscle of normal and dystrophic mice, *IRCS Med. Sci.*, 7, 136, 1979.
62. **Beitner, R., Nordenberg, J., and Cohen, T. J.,** Correlation between the levels of glucose-1,6-diphosphate and the activities of phosphofructokinase, phosphoglucomutase and hexokinase, in skeletal and heart muscles from rats of different ages, *Int. J. Biochem.*, 10, 603, 1979.
63. **Rowland, L. P.,** Pathogenesis of muscular dystrophies, *Arch. Neurol.*, 33, 315, 1976.
64. **Strickland, K. P., Hudson, A. J., and Thakar, J. H.,** Biochemical studies in dystrophic mouse muscle, *Ann. N.Y. Acad. Sci.*, 317, 187, 1979.
65. **Kwock, C. T. and Austin, L.,** Phospholipid composition and metabolism in mouse muscular dystrophy, *Biochem. J.*, 176, 15, 1978.
66. **Kunze, D.,** Lipids: composition and metabolism in human dystrophy, in *Pathogenesis of Human Muscular Dystrophies*, Rowland, L. P., Ed., Excerpta Medica, Amsterdam, 1977, 404.
67. **Jain, M. K.,** Enzymic hydrolysis of various components in biomembranes and related systems, in *Current Topics in Membranes and Transport*, Vol. 4, Bronner, F. and Kleinzeller, A., Eds., Academic Press, New York, 1973, 175.
68. **Meldrum, B. S.,** The actions of snake venoms on nerve and muscle. The pharmacology of phospholipase A and of polypeptide toxins, *Pharm. Rev.*, 17, 393, 1965.
69. **Chain, E.,** Effect of snake venoms on glycolysis and termentation in cell-free extracts, *Q. J. Exp. Physiol.*, 26, 299, 1937.

70. **Chain, E.,** Neutralisation of the antifermenting principle in snake venoms by antivenins, *Q. J. Exp. Physiol.*, 27, 49, 1938.
71. **Reid, H. A.,** Diagnosis, prognosis and treatment of sea-snake bite, *Lancet*, 2, 399, 1961.
72. **Reid, H. A.,** Myoglobinuria and sea-snake bite poisoning, *Br. Med. J.*, 1, 1284, 1961.
73. **Beitner, R., Cohen, T. J., Nordenberg, J., and Haberman, S.,** Glucose-1,6-bisphosphate and the mechanism of Pasteur effect in diaphragm muscle. The regulation of hexokinase and phosphofructokinase activities, *Biochim. Biophys. Acta*, 586, 266, 1979.
74. **Passonneau, J. V., Lowry, O. H., Schulz, D. W., and Brown, J. G.,** Glucose 1,6-diphosphate formation by phosphoglucomutase in mammalian tissues, *J. Biol. Chem.*, 244, 902, 1969.
75. **Ramaiah, A.,** Pasteur effect and phosphofructokinase, in *Current Topics in Cellular Regulation*, Vol. 8, Horecker, B. L. and Stadtman, E. R., Eds., Academic Press, New York, 1974, 297.
76. **Tejwani, G. A.,** The role of phosphofructokinase in Pasteur effect, *Trends Biochem. Sci.*, 3, 30, 1978.
77. **Tejwani, G. A., Ramaiah, A., and Ananthanarayanan, M.,** Regulation of glycolysis in muscle. The role of ammonium and synergism among the positive effectors of phosphofructokinase, *Arch. Biochem. Biophys.*, 158, 195, 1973.
78. **Ramaiah, A.,** Regulation of glycolysis in skeletal muscle, *Life Sci.*, 19, 455, 1976.
79. **Schudt, C. and Pette, D.,** Ca^{2+}-ions as coupling agents in enzymatic differentiation and carbohydrate metabolism in cultured muscle cells, in *Advances in Enzyme Regulation*, Vol. 16, Weber, G., Ed., Pergamon Press, Oxford, 1978, 121.
80. **Beitner, R., Klein, S., and Nordenberg, J.,** The participation of glucose-1,6-diphosphate in the regulation of hexokinase and phosphoglucomutase activities in brains of young and adult rats, *Int. J. Biochem.*, 14, 195, 1982.
81. **Baquer, N. Z., McLean, P., and Greenbaum, A. L.,** Enzymatic differentiation in pathways of carbohydrate metabolism in developing brain, *Biochem. Biophys. Res. Commun.*, 53, 1282, 1973.
82. **Nordenberg, J., Heffetz, D., Cohen, T. J., and Beitner, R.,** Glucose-1,6-diphosphate and carbohydrate metabolism in skeletal muscle of old rats, *Int. J. Biochem.*, 13, 317, 1981.
83. **Ermini, M., Shelenyi, I., and Moser, P.,** Altersbedingte abnahme von kreatinphosphat und änderungen der adeninnukleotide in der skelettmuskulatur von ratten, *Experientia*, 28, 401, 1972.
84. **Ermini, M.,** Aging changes in mammalian skeletal muscle, *Gerontologia*, 22, 301, 1976.
85. **Pieri, C. and Ermini, M.,** Changes of fibre pattern in the diaphragmatic muscle of aging rats, *Experientia*, 30, 215, 1974.
86. **Benoit, P. W. and Belt, W. D.,** Some effects of local anesthetic agents on skeletal muscle, *Exp. Neurol.*, 34, 264, 1972.
87. **Yagiela, J. A. and Benoit, P. W.,** Skeletal muscle damage from quinidine, *N. Engl. J. Med.*, 301, 437, 1979.
88. **Foster, A. M. and Carlson, B. M.,** Myotoxicity of local anesthetics and regeneration of the damaged muscle fibers, *Anesth. Analg.*, 59, 727, 1980.
89. **Reed, P. W. and Lardy, H. A.,** A23187: a divalent cation ionophore, *J. Biol. Chem.*, 247, 6970, 1972.
90. **Publicover, S. J., Duncan, C. J., and Smith, J. L.,** The use of A23187 to demonstrate the role of intracellular calcium in causing ultrastructural damage in mammalian muscle, *J. Neuropathol. Exp. Neurol.*, 37, 544, 1978.
91. **Rose, I. A. and Warms, J. V. B.,** Lithium action on glucose uptake in brain; role of Glc-1,6-P_2 as a regulator of hexokinase, *Biochem. Biophys. Res. Commun.*, 92, 1030, 1980.
92. **Nordenberg, J., Kaplansky, M., Beery, E., Klein, S., and Beitner, R.,** Effects of lithium on the activities of phosphofructokinase and phosphoglucomutase and on glucose-1,6-diphosphate levels in rat muscles, brain and liver, *Biochem. Pharmacol.*, 31, 1025, 1982.
93. **Nordenberg, J., Kaplansky, M., and Beitner, R.,** Influence of lithium on glucose-1,6-diphosphate levels in muscle of normal and dystrophic mice, *IRCS Med. Sci.*, 8, 880, 1980.
94. **Frucht, H., Kaplansky, M., and Beitner, R.,** Increase in glucose 1,6-bisphosphate levels, activation of phosphofructokinase and phosphoglucomutase and inhibition of glucose 1,6-bisphosphatase in muscle induced by trifluoperazine, *Biochem. Med.*, 31, 122, 1984.
95. **Sener, A., Malaisse-Lagae, F., and Malaisse, W. J.,** Glucose-induced accumulation of glucose-1,6-bisphosphate in pancreatic islets: its possible role in the regulation of glycolysis, *Biochem. Biophys. Res. Commun.*, 104, 1033, 1982.
96. **Rose, I. A., Warms, J. V. B., and Kaklij, G.,** A specific enzyme for glucose-1,6-bisphosphate synthesis, *J. Biol. Chem.*, 250, 3466, 1975.
97. **Wong, L.-J. and Rose, I. A.,** Kinetic competence of a phosphoryl enzyme intermediate in the Glc-1,6-P_2 synthase-catalyzed reaction, *J. Biol. Chem.*, 251, 5431, 1976.
98. **Alpers, J. B. and Lam, G. K. H.,** Studies on phosphomutases. Phosphoryl transfer to phosphoglucomutase from 1,3-diphosphoglycerate, *J. Biol. Chem.*, 244, 200, 1969.

99. **Eyer, P. and Pette, D.,** Purification and properties of "glucose-1-phosphate kinase", *Life Sci.*, 6, 191, 1967.
100. **Eyer, P., Hofer, H. W., Krystek, E., and Pette, D.,** Synthesis of glucose-1,6-bisphosphate by the action of crystalline rabbit muscle phosphofructokinase, *Eur. J. Biochem.*, 20, 153, 1971.
101. **Paladini, A. C., Caputto, R., Leloir, L. F., Trucco, R. E., and Cardini, C. E.,** The enzymatic synthesis of glucose-1,6-diphosphate, *Arch. Biochem.*, 23, 55, 1949.
102. **Sidbury, J. B., Rosenberg, L. L., and Najjar, V. A.,** Muscle glucose-1-phosphate transphosphorylase, *J. Biol. Chem.*, 222, 89, 1956.
103. **Hashimoto, T. and Yoshikawa, H.,** The presence of D-glucose 1,6-diphosphate hydrolyzing enzymes, *J. Biochem. (Tokyo)*, 59, 427, 1966.
104. **Hashimoto, T. and Yoshikawa, H.,** Purification and properties of liver acid phosphatase hydrolyzing glucose-1,6-diphosphate, *J. Biochem. (Tokyo)*, 61, 760, 1967.
105. **Beitner, R. and Cohen, T. J.,** An increase in the activity of glucose-1,6-diphosphate phosphatase in muscle of dystrophic mice, *IRCS Med. Sci.*, 7, 24, 1979.
106. **Wakelam, M. J. O., Emmerich, M., and Pette, D.,** The control of glucose-1,6-bisphosphatase by Ca^{2+} and calmodulin, *Biochem. J.*, 208, 517, 1982.
107. **Guha, S. K. and Rose, Z. B.,** Brain glucose bisphosphatase requires inosine monophosphate, *J. Biol. Chem.*, 257, 6634, 1982.
108. **Goldberg, N. D., Haddox, M. K., Nicol, S. E., Glass, D. B., Sanford, C. H., Kuehl, F. A., Jr., and Estensen, R.,** Biologic regulation through opposing influences of cyclic GMP and cyclic AMP: the Yin-Yang hypothesis, in *Advances in Cyclic Nucleotide Research*, Vol. 5, Drummond, G. I., Greengard, P., and Robison, G. A., Eds., Raven Press, New York, 1975, 307.
109. **Goldberg, N. D. and Haddox, M. K.,** Cyclic GMP metabolism and involvement in biological regulation, *Annu. Rev. Biochem.*, 46, 823, 1977.
110. **Wong, P. V.-K. and Cheung, W. Y.,** Calmodulin stimulates human platelet phospholipase A_2, *Biochem. Biophys. Res. Commun.*, 30, 473, 1979.
111. **Cheung, W. Y.,** Calmodulin plays a pivotal role in cellular regulation, *Science*, 207, 19, 1980.
112. **Blosser, J. C. and Appel, S. H.,** Properties and distribution of mammalian skeletal muscle guanylate cyclase, *J. Biol. Chem.*, 253, 3088, 1978.
113. **Schier, W. T., Baldwin, J. H., Nilsen-Hamilton, M., Hamilton, R. T., and Thanassi, N. M.,** Regulation of guanylate and adenylate cyclase activities by lysolecithin, *Proc. Natl. Acad. Sci. U.S.A.*, 73, 1586, 1976.
114. **Sulakhe, S. J., Leung, N. L.-K., and Sulakhe, P. V.,** Properties of particulate, membrane-associated and soluble guanylate cyclase from cardiac muscle, skeletal muscle, cerebral cortex and liver, *Biochem. J.*, 157, 713, 1976.
115. **Fujimoto, M. and Okabayashi, T.,** Proposed mechanisms of stimulation and inhibition of guanylate cyclase with reference to the actions of chlorpromazine phospholipases and Triton X-100, *Biochem. Biophys. Res. Commun.*, 67, 1332, 1975.
116. **Beitner, R., Lilling, G., Frucht, H., Ben-Porat, H., and Sofer, Y.,** Age-dependent changes in glucose-1,6-bisphosphate levels and in the activities of glucose-1,6-bisphosphatase, and particulate hexokinase and 6-phosphogluconate dehydrogenase in rat skin, *Biochem. Med.*, 30, 369, 1983.
117. **Ziboh, V. A. and Lord, J. T.,** Phospholipase A activity in the skin. Modulators of arachidonic acid release from phosphatidylcholine, *Biochem. J.*, 184, 283, 1979.
118. **Hers, H. G., Hue, L., and Van Schaftingen, E.,** Fructose 2,6-bis-phosphate, *Trends Biochem. Sci.*, 7, 329, 1982.
119. **Pilkis, S. J., El-Maghrabi, M. R., McGrane, M., Pilkis, J., Fox, E., and Claus, T. H.,** Fructose 2,6-bisphosphate: a mediator of hormone action at the fructose 6-bisphosphate/fructose 1,6-bisphosphate substrate cycle, *Mol. Cell. Endocrinol.*, 25, 245, 1982.
120. **Hers, H. G. and Van Schaftingen, E.,** Fructose 2,6-bisphosphate 2 years after its discovery, *Biochem. J.*, 206, 1, 1982.
121. **Kuwajima, M. and Uyeda, K.,** The tissue distribution of fructose-2,6-P_2 and fructose-6-P, 2-kinase in rats and the effect of starvation diabetes and hypoglycemia on hepatic fructose- 2,6-P_2 and fructose-6-P, 2-kinase, *Biochem. Biophys. Res. Commun.*, 104, 84, 1982.
122. **Bessman, S. P. and Geiger, P. J.,** Compartmentation of hexokinase and creatine phosphokinase, cellular regulation, and insulin action, in *Current Topics in Cellular Regulation*, Vol. 16, Horecker, B. L. and Stadtman, E. R., Eds., Academic Press, London, 1980, 55.
123. **Beitner, R.,** unpublished results.

Chapter 2

ROLE OF FRUCTOSE 2,6-BISPHOSPHATE IN THE CONTROL OF GLYCOLYSIS IN LIVER, MUSCLE, AND ADIPOSE TISSUE

Louis Hue and Ramon Bartrons

TABLE OF CONTENTS

I. INTRODUCTION

Fructose-2,6-bisphosphate was discovered[1] during the study of the action of glucagon on the fructose 6-phosphate/fructose-1,6-bisphosphate cycle. In the liver, glucagon inhibits glycolysis and stimulates gluconeogenesis. This hormone has been shown to change the kinetic properties of phosphofructokinase resulting in a decreased affinity for fructose 6-phosphate and an increased inhibition by ATP and citrate without a change in V_{max}.[2-5] These modifications suggested that, like pyruvate kinase, phosphofructokinase could be regulated by covalent modification of the protein. Glucagon, through the second messenger cyclic AMP, activates cyclic AMP-dependent protein kinase which, in turn, catalyzes the phosphorylation of various proteins. Phosphofructokinase could conceivably be one of these proteins. However, subsequent studies showed that the difference in phosphofructokinase activity could also be obtained by gel filtration of an extract from control hepatocytes.[6] Moreover, the effect could be abolished by adding a low-molecular-weight fraction obtained from control cells, but not from glucagon-treated cells to the purified preparation.[6] This stimulatory factor was purified and its properties indicated that it was a nonnucleotidic-diphosphoric ester, which was extremely acid labile. Hydrolysis of the partially purified stimulator yielded equimolar amounts of fructose 6-phosphate, P_i, and a reducing group suggesting that phosphate was bound to the second carbon of fructose 6-phosphate as in fructose-2,6-bisphosphate.[1]

The synthesis of fructose-2,6-bisphosphate from fructose 6-phosphate and phosphoric acid by reversal of the acid hydrolysis unequivocally demonstrated that the stimulator was indeed fructose-2,6-bisphosphate.[7] Later fructose-2,6-bisphosphate was synthesized on a large scale by conversion of fructose-1,6-bisphosphate into the 1,2 cyclic-phosphodiester and alkaline hydrolysis of the latter.[8-10] NMR analysis showed that it was the β stereoisomer.[9-11]

In an endeavor to define the role of fructose-2,6-bisphosphate in the regulation of metabolism, one must take into consideration its effect in vitro on enzyme activities as well as the factors and conditions which modulate its concentration in tissues. Fructose-2,6-bisphosphate has been detected in liver, brain, kidney, heart, skeletal muscle, adipose tissue, lung, and pancreatic islets. Fructose-2,6-bisphosphate has also been found in yeasts and plants. This survey is limited to mammalian tissues and in particular to liver, skeletal muscle, and adipose tissue since very little, if anything, is known about fructose-2,6-bisphosphate in other tissues. Fructose-2,6-bisphosphate has been the subject of several recent reviews.[12-16]

II. CONTROL OF ENZYME ACTIVITY BY FRUCTOSE-2,6-BISPHOSPHATE

A. Phosphofructokinase

Fructose-2,6-bisphosphate is the most potent known stimulator of phosphofructokinase.[9,17,18] It affects the activity of all mammalian phosphofructokinases studied so far. The kinetic characteristics of phosphofructokinase are such that ATP, one of its substrates, behaves as a negative allosteric effector inducing cooperativity for fructose 6-phosphate, the second substrate. Fructose 6-phosphate acts as a positive effector, relieving the inhibition by ATP. Several other allosteric effectors are known which exert an effect similar to, and synergistic with, that of ATP (negative effector) or fructose 6-phosphate (positive effector). Fructose-1,6-bisphosphate, fructose-2,6-bisphosphate, glucose-1,6-bisphosphate, AMP, P_i, NH_4^+, and 6-phosphoguconate are positive effectors whereas citrate and sn-glycerol 3-phosphate are negative effectors. In the presence of physiological concentrations of fructose 6-phosphate, ATP, AMP, P_i, and NH_4^+, the activity of liver phosphofructokinase is almost negligible and physiological concentrations of fructose 2,6-bisphosphate (in the micromolar range) relieve the inhibition by ATP and allow glycolysis to proceed.[17,18] Fructose-2,6-

bisphosphate is, therefore, a good candidate for the control of glycolysis under plethoric conditions, when energy is plentiful and when glycolysis should proceed for biosynthetic purposes such as the provision of C3 units for lipogenesis. In contrast, when the supply of ATP by glycolysis becomes important, adenine nucleotides and P_i are likely to play the major role. From these kinetic studies it also appeared that, at least in the liver, fructose-1,6-bisphosphate and glucose-1,6-bisphosphate are unlikely to exert any significant stimulatory effect under normal conditions since their concentration in the cells are too low to affect phosphofructokinase activity. This may be not valid for other tissues such as the skeletal muscle in which the concentration of fructose-1,6-bisphosphate can reach millimolar values.[19]

B. Fructose-1,6-Bisphosphatase

Fructose-2,6-bisphosphate is also an inhibitor of fructose-1,6-bisphosphatase.[20-23] This enzyme displays a high affinity for its substrate (K_m in the micromolar range) and is inhibited allosterically by AMP. The inhibition by fructose-2,6-bisphosphate is synergistic with AMP and stronger at low than at high substrate concentration. Fructose-2,6-bisphosphate changes the saturation curve for the substrate from a hyperbolic to a sigmoidal shape suggesting an allosteric type of interaction. From recent kinetic studies, it appears that the action of fructose-2,6-bisphosphate is complex and probably involves its binding to a specific site which is different from the site for fructose-1,6-bisphosphate;[24] however, some overlapping between the two sites is not excluded. Binding studies have shown that fructose-1,6–bisphosphate binds 1 mol of fructose-2,6-bisphosphate per mole of subunit.[25] The V_{max} of rat liver fructose-1,6-bisphosphatase is several fold higher than that of phosphofructokinase. Therefore when the metabolism of the liver changes such that lactate is produced rather than consumed, fructose-1,6-bisphosphatase should be strongly inhibited. Physiological concentrations of AMP (total cellular concentration around 0.1 to 0.2 m*M*[26]; see, however, below) are high enough to completely inhibit fructose-1,6-bisphosphatase and fructose-2,6-bisphosphate further reinforces this inhibition.

However, in the livers from fed rats containing about 10 μ*M* fructose-2,6-bisphosphate, recycling of metabolites between glucose and triose phosphates occurs indicating that both phosphofructokinase and fructose-1,6-bisphosphatase are active.[6,26] Therefore this suggests that the inhibition of fructose-1,6-bisphosphatase in vivo is less than expected from in vitro measurements and that fructose-1,6-bisphosphatase is less sensitive to fructose-2,6-bisphosphate than phosphofructokinase. The latter proposal is in agreement with kinetic[17,18,21,23] and binding[24] studies which showed that the affinity constant of phosphofructokinase for fructose-2,6-bisphosphate is about one order of magnitude greater than that of fructose-1,6-bisphosphatase.

The total cytosolic AMP concentration is about 100 μ*M*[27] and the concentration of AMP-binding sites in the cytosol is far from negligible.[28] Therefore the concentration of free AMP in the cytosol might well be in the micromolar range. Variations in both AMP and fructose-2,6-bisphosphate are therefore crucial events in the regulation of the activity of phosphofructokinase and fructose-1,6-bisphosphatase.

Finally it should be added that rat liver fructose-1,6-bisphosphatase and phosphofructokinase are substrates of cyclic AMP-dependent protein kinase but that phosphorylation causes little, if any change in their kinetic properties.[16]

C. Effects on Other Enzymes

In mammalian tissues, the only known effect of fructose-2,6-bisphosphate is at the level of phosphofructokinase/fructose-1,6-bisphosphatase. No other effect has been described so far. On the other hand, fructose-2,6-bisphosphate is a stimulator of PP_i: fructose 6-phosphate phosphotransferase, an enzyme which is present in higher plants;[29] however, the role of this enzyme in plant metabolism is little understood. Because of its extreme sensitivity towards fructose-2,6-bisphosphate, the PP_i phosphofructokinase from potato tubers is currently used to measure fructose-2,6-bisphosphate in tissue extracts.[30] Fructose-2,6-bisphosphate has also been shown to be an inhibitor of UDPglucose phosphorylase from potato tubers.[31] Again, the physiological significance of this effect is not clear.

III. SYNTHESIS AND BREAKDOWN OF FRUCTOSE-2,6-BISPHOSPHATE

The enzymes responsible for the synthesis and breakdown of fructose-2,6-bisphosphate have only been studied in the liver. Fructose-2,6-bisphosphate is synthesized from fructose 6-phosphate and ATP by 6-phosphofructo-2-kinase, also called phosphofructokinase-2.[32-35] In the presence of 5 mM P_i, the K_m for fructose 6-phosphate is 50 μM; i.e., in the physiological range. The activity of the enzyme is inhibited by physiological concentrations of citrate, phosphoenolpyruvate, and sn-glycerol 3-phosphate. Fructose-2,6-bisphosphate is hydrolyzed to fructose 6-phosphate and P_i by a phosphatase called fructose-2,6-bisphosphatase.[36-38] The K_m for fructose-2,6-bisphosphate is very low (below micromolar;[37] however much higher values have been reported[36,38]) and much smaller than the usual concentrations of fructose-2,6-bisphosphate in the livers of fed animals. The enzyme is strongly inhibited by fructose 6-phosphate (Ki: 10 to 40 μM)[37,38] and is stimulated by physiological concentrations of P_i,[37,38] sn-glycerol 3-phosphate,[37] and nucleotide triphosphates.[37] The V_{max} of each enzyme is relatively low (less than 20 mU/g of liver) and changes in fructose-2,6-bisphosphate concentration require several minutes to become complete. In addition, both liver enzymes are substrates for cyclic AMP-dependent protein kinase.[35-42] Phosphorylation of the protein (1 mol of P_i per mole of subunit) results in an inactivation of phosphofructokinase-2 and an activaton of fructose-2,6-bisphosphatase.

Various purification and fractionation procedures have not allowed the separation of phosphofructokinase-2 from fructose-2,6-bisphosphatase,[37-40,43] suggesting that the two catalytic properties belong to a single protein. It has been recently demonstrated that the enzyme purified to homogeneity catalyzes both phosphohydrolase and phosphotransferase reactions and that a histidyl residue of the protein becomes phosphorylated in the presence of fructose-2,6-bisphosphate or ATP plus fructose 6-phosphate.[43]

The fact that phosphofructokinase-2/fructose-2,6-bisphosphatase is sensitive to various metabolites and regulated in a coordinated fashion offers the basis for integration and concentration of information on a single control point, fructose-2,6-bisphosphate. Moreover fructose-2,6-bisphosphate may express information from other metabolic pathways at the level of phosphofructokinase and fructose-1,6-bisphosphatase.

IV. ROLE OF FRUCTOSE-2,6-BISPHOSPHATE IN THE LIVER

Since fructose-2,6-bisphosphate affects the activity of both liver phosphofructokinase and fructose-1,6-bisphosphatase, its most obvious role is the control of glycolysis and gluconeogenesis. The flux through these pathways is under hormonal and nonhormonal control and depends on the nutritional status of the animal.

The liver can be in three metabolic states. In the first situation, net synthesis of glucose occurs and glycolysis should be stopped to avoid futile cycling. In the second case, glucose is in excess, and glycolysis operates for biosynthetic purposes such as the provision of C3 units for the *de novo* synthesis of fatty acids. Finally, glycolysis can be stimulated during anoxia when it serves the purpose of supplying ATP. These different conditions will be reviewed and for each the role of fructose-2,6-bisphosphate will be considered.

A. Effect of Glucagon and Insulin

Glucagon is known to stimulate gluconeogenesis and to arrest glycolysis. The hormone causes a fall in the concentration of fructose-2,6-bisphosphate which can be explained by the inactivation of phosphofructokinase-2 and the activation of fructose-2,6-bisphosphatase.[35-42] These changes result from the phosphorylation of the enzyme by cyclic AMP-dependent protein kinase.

We found that the effects of glucagon to affect the fructose-2,6-bisphosphate system are obtained at the same doses that stimulate glycogen breakdown and the inactivation of pyruvate kinase[35,44] (Figure 1). Differences in sensitivity have however been reported by others.[38,45-47] In addition, the fructose-2,6-bisphosphate system is slower to react after glucagon treatment than pyruvate kinase and phosphorylase,[44] suggesting that phosphofructokinase-2/fructose-2,6-bisphosphatase are relatively poor substrates for the cyclic AMP-dependent protein kinase.

The fall in fructose-2,6-bisphosphate brought about by glucagon increases fructose-1,6-bisphosphatase activity, decreases phosphofructokinase activity and eventually stops the flux through this enzyme. Since glucagon also causes the inactivation of pyruvate kinase,[48] the whole glycolytic flux is blocked by this dual lock. The extent of the inhibition of glycolysis by glucagon, however, depends on the presence of glycogen. In the fed state, glucagon promotes glycogen breakdown and causes an increase in the concentration of hexose 6-phosphates which stimulates phosphofructokinase-1 and -2, thus counteracting the effect of glucagon on phosphofructokinase-2/fructose-2,6-bisphosphatase. Indeed, the experimental evidence shows that in the livers of fed rats treated with glucagon, glycolytic flux is not entirely blocked and that recycling occurs between glucose and triose phosphates.[6,26] In livers of starved rats such a situation does not exist, phosphofructokinase-1 and -2 are inactive and recycling between fructose 6-phosphate and fructose-1,6-bisphosphate cannot be measured.[6,26]

It is questionable whether a fall in fructose-2,6-bisphosphate is an obligatory step in the stimulation of gluconeogenesis by glucagon. From recent studies,[49] it appears that the physiological importance of the two effects of glucagon of inactivating pyruvate kinase and decreasing fructose-2,6-bisphosphate depends on the nature of the substrates. With substrates entering the pathway at the level of pyruvate, the inactivation of pyruvate kinase appears essential in redirecting phosphoenolpyruvate towards glucose, whereas with substrates entering at the level of triose phosphates, a decrease in fructose-2,6-bisphosphate seems more important.

Insulin may counteract the increase in cyclic AMP,[50] the inactivation of pyruvate kinase,[51,52] and phosphofructokinase brought about by suboptimal doses of glucagon.[2] The latter effect results from changes in the concentration of fructose-2,6-bisphosphate.[53]

B. Effect of Adenosine

Adenosine and its analog, 2-chloroadenosine, have been shown to decrease the concentration of fructose-2,6-bisphosphate in rat hepatocytes incubated with glucose.[54] This effect was accompanied by a rise in cyclic AMP, causing an activation of phosphorylase and fructose-2,6-bisphosphatase and an inactivation of pyruvate kinase and phosphofructokinase-2. Since the effect of the nucleoside was not abolished by inhibitors of the adenosine transport system, it was suggested that the binding of adenosine to its receptors caused a stimulation of adenylate cyclase. β-Adrenergic agonists, which act through cyclic AMP-dependent mechanisms, have the same effect as glucagon and adenosine.

C. Alpha-Adrenergic Agents and Vasopressin

In contrast to the generally accepted view, α-adrenergic agents and vasopressin do not always stimulate liver gluconeogenesis. Although there is little doubt that these agents

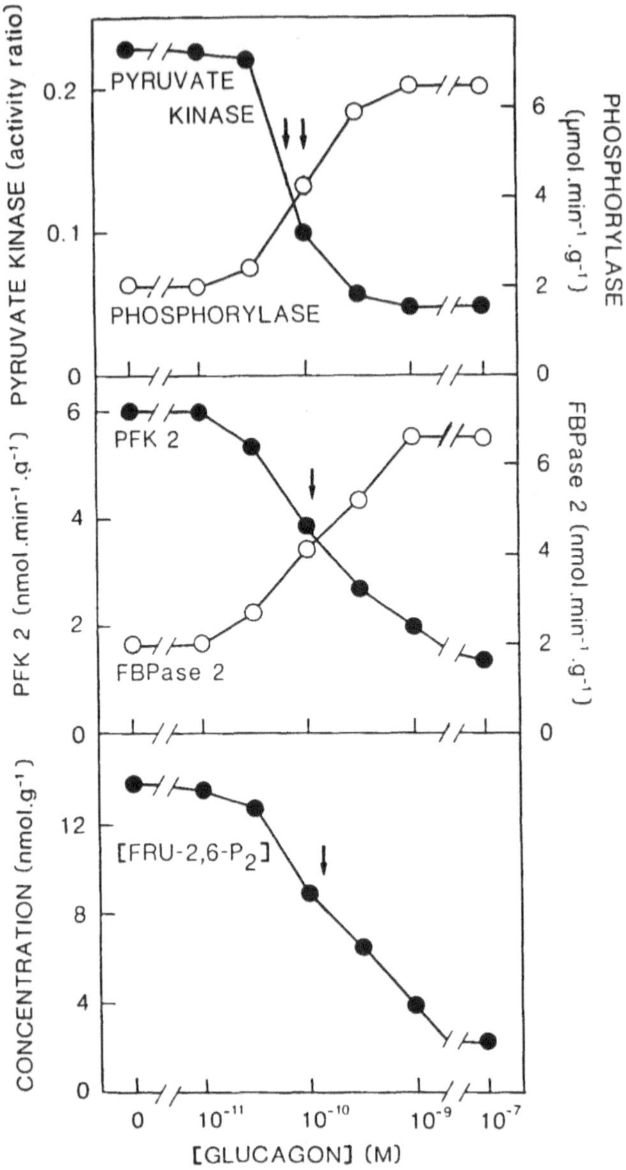

FIGURE 1. Dose-response curves of the effect of glucagon on the interconversion of enzymes, and on the concentration of fructose-2,6-bisphosphate. Hepatocytes from a fed rat were incubated for 6 min at 37°C in the presence of various concentrations of glucagon. Arrows indicate the concentration of hormone causing a half-maximal effect. (From Bartrons, R., Hue, L., Van Schaftingen, E., and Hers, H. G., *Biochem. J.*, 214, 829, 1983. With permission.)

stimulate gluconeogenesis in the starved state, in the fed state, however, they increase both glycogenolysis and glycolysis.[35,55] They stimulate glycogen breakdown by a mechanism independent of cyclic AMP and probably mediated by Ca^{++},[56,57] which causes an increase in fructose-2,6-bisphosphate resulting from the accumulation of hexose 6-phosphate.[35] This increase is not counteracted, as in the glucagon-treated livers, by an inactivation of phosphofructokinase-2 and an activation of fructose-2,6-bisphosphatase. Since α-adrenergic ag-

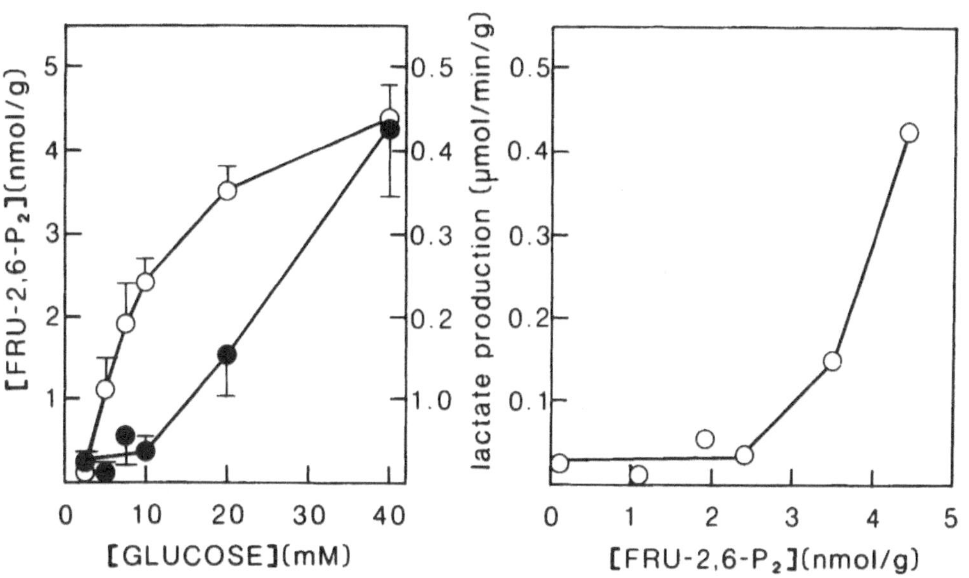

FIGURE 2. Relationship between lactate production and the concentration of fructose-2,6-bisphosphate in hepatocytes incubated with glucose. Hepatocytes from overnight-starved rats were incubated for 30 min at 37°C in the presence of various concentrations of glucose.

onists and vasopressin have little effect on pyruvate kinase,[58] they cause an increase in lactate output. Because vasopressin and phenylephrine also activate pyruvate dehydrogenase[59,60] and stimulate recycling at the level of pyruvate/phosphoenolpyruvate,[6] the stimulation of lactate output is less than expected from the increase in fructose-2,6-bisphosphate concentration. This glycogen-dependent mechanism offers an explanation for the stimulation of glycolysis by α-adrenergic agonists and vasopressin.

There has been report claiming that adrenaline caused a Ca^{++}-dependent inactivation of phosphofructokinase-2 and activation of fructose-2,6-bisphosphatase.[46,47] This has not been confirmed[44] and the discrepancy is difficult to interpret; however, it should be noted that the response of rat hepatocytes to adrenaline is known to depend on the age[62] and the sex[63] of the animals.

D. Glucose and Other Glycolytic Substrates

In addition to glycogen, glucose is another substrate for liver glycolysis. In perfused livers from fed rats, the relationship between lactate production and the concentration of glucose is not linear:[64,65] addition of 5 to 10 m*M* glucose has little effect whereas concentrations larger than 20 m*M* greatly increase the production of lactate. In isolated hepatocytes from overnight-starved rats, a similar response is observed (Figure 2). Figure 2 also shows that the relationship between the concentration of fructose-2,6-bisphosphate and the production of lactate is not linear: there are conditions under which an increase in fructose-2,6-bisphosphate is not related to a stimulation of lactate output. A similar situation is observed when hepatocytes are incubated with low concentrations of dihydroxyacetone or fructose.[49] At concentrations less than 2 m*M*, these substrates cause an increase in fructose-2,6-bisphosphate, but the output of lactate remains low. Furthermore, the data strongly suggest that, under these conditions, there is an increase in cycling between fructose 6-phosphate and fructose-1,6-bisphosphate.[49] Therefore it seems likely that, phosphofructokinase is more sensitive to fructose-2,6-bisphosphate than fructose-1,6-bisphosphatase. In other words, at low concentrations of fructose-2,6-bisphosphate, both phosphofructokinase and fructose-1,6-bisphosphatase are active causing a futile cycle. Since the V_{max} of fructose-1,6-bisphosphatase is much larger than that of phosphofructokinase,[16] the net flux is probably oriented towards glucose formation. On the other hand, when the concentration of fructose-2,6-bisphosphate is further increased, fructose-1,6-bisphosphatase becomes progressively inhibited causing net glycolysis to occur. However, the inhibition is not complete and cycling persists.

Therefore, the metabolic significance of fructose-2,6-bisphosphate is best expressed at relatively high concentrations when it allows the pathway to switch from gluconeogenesis to glycolysis and vice versa. If glucose is in excess, the liver stores some as glycogen and transforms the excess into fatty acids.

Glycogen metabolism in the liver represents another system which is glucose sensitive with phosphorylase acting as a glucose sensor.[66,67] When enough glucose is given, the liver stops producing glucose and starts to store it as glycogen; the concentration of glucose at which this occurs is sometimes referred to as the "threshold" for glucose. It would be interesting to know whether the threshold is the same for glycolysis and glycogen synthesis. Caution should however be exercised in applying conclusions drawn from experiments done with hepatocytes to the situation in vivo since the glucose sensitivity in vitro is not necessarily the same as in vivo.[68]

E. Anoxia

The glycolytic rate can be greatly increased by anoxia. This phenomenon, often referred to as the Pasteur effect,[69-71] can be demonstrated in almost all tissues and is of particular importance in skeletal muscle for the provision of energy during violent anaerobic contractions.[72-73] The mechanisms responsible for the acceleration of the rate are generally believed to be changes in adenine nucleotides causing a stimulation of phosphofructokinase activity. The role of fructose-2,6-bisphosphate in this process has been studied in isolated hepatocytes. When glycolysis was stimulated by anoxia, the concentrations of fructose-2,6-bisphosphate decreased indicating that this ester could not be responsible for the onset of anaerobic glycolysis.[74] In addition, the effect of glucose to increase fructose-2,6-bisphosphate concentrations under aerobic conditions was greatly impaired in anoxic hepatocytes. The role of fructose-2,6-bisphosphate in the stimulation of liver glycolysis is therefore restricted to plethoric conditions when glucose is in excess. Under anaerobic or ischemic conditions, the predominant role is probably not exerted by fructose-2,6-bisphosphate but by adenine nucleotides, P_i, and citrate.

F. Gluconeogenic Substrates

Incubation of hepatocytes from fed rats with relatively high concentrations (20 m*M*) of keto-hexoses such as D-fructose, D-tagatose, and L-sorbose caused a drastic decrease in fructose-2,6-bisphosphate.[75] In the presence of high concentrations of glyceraldehyde and glycerol, but not dihydroxyacetone, fructose-2,6-bisphosphate was also greatly reduced. Except dihydroxyacetone, all these compounds have the common property of causing a rapid accumulation of ester phosphates and, in so doing, deplete the adenine nucleotides and the P_i pools of the cell.[76] Incubations of hepatocytes with 20 m*M* lactate also caused a fall in fructose-2,6-bisphosphate which became comparable to values obtained after glucagon treatment. Pyruvate and alanine had a less marked effect.[75] The mechanism by which lactate, and to a lesser extent alanine and pyruvate, decrease fructose-2,6-bisphosphate is not known. This effect may result from an inhibition of phosphofructokinase-2 by phosphoenolpyruvate, which accumulates in livers perfused with these substrates.[50] As a result of the decrease in fructose-2,6-bisphosphate, gluconeogenesis is favored and the glycolytic flux is inhibited at the level of phosphofructokinase.

Incubation of hepatocytes from fasted rats with dihydroxyacetone, and low concentrations of fructose, caused an increase in fructose-2,6-bisphosphate. Lactate and alanine had a small effect whereas glycerol had no effect.[49]

On the other hand, glucose may inhibit gluconeogenesis from alanine, pyruvate, and lactate.[52,77,78] This inhibition may be related to an increase in fructose-2,6-bisphosphate.

Finally, it should be mentioned that ethanol is another substance which causes a fall in fructose-2,6-bisphosphate in liver.[79]

G. Long-Term Adaptation and Obesity

During long-term adaptation to fasting or diabetes many changes take place in the liver. Some of these changes are concerned with the regulation of gluconeogenesis: in livers of starved or diabetic rats, the concentration of cyclic AMP is increased,[52] the total activity as well as the proportion of pyruvate kinase and phosphofructokinase in the active form is decreased,[16,80-83] and the concentration of fructose-2,6-bisphosphate is very low.[84,85] All these modifications contribute to the maintenance of a high gluconeogenic state. Fructose-2,6-bisphosphate has many properties in common with the phosphofructokinase-stabilizing factor[86] described by Dunaway[87,88] and its disappearance during fasting and diabetes[83] could account for the faster rate of degradation of phosphofructokinase which has been reported to occur under these conditions.

Conversely, in livers of genetically obese mice[89,90] and rats,[91] the concentration of fructose-2,6-bisphosphate is larger than in livers of lean controls. This is particularly striking in livers of starved obese animals which retain a considerable amount of glycogen, produce lactate, and display detectable rates of lipogenesis.[92] Thus the metabolic state of livers from starved obese rats is similar to that of normal fed rats but quite opposite to that of normal starved rats. These changes may be explained in terms of cyclic AMP which was found to be decreased in the livers of starved obese rats.[91,93] In livers of fed obese rats, a decrease in cyclic AMP was not observed and the high fructose-2,6-bisphosphate level may result from a long-term adaptation such as an increase in phosphofructokinase-2 activity.[91]

V. ROLE IN MUSCLE TISSUE

In skeletal muscle glycolysis is stimulated by contraction and by treatment with hormones such as adrenaline and insulin.[73,94] During contraction, the extent of stimulation of glycolysis depends on the frequency of stimulation/contraction and the presence of oxygen, two conditions which are interconnected. Two extreme situations can be observed: a series of twitches occurring at frequency lower than 5/sec causing little increase in glycolysis, and, on the other hand, tetanization of the muscle when the frequency of stimulation is higher than 20/sec. This latter condition results in a maximal stimulation of glycolysis, and the muscle very rapidly reaches a state of fatigue in which the tension developed gradually decreases.

Electrical stimulation (10 pulses/sec) of perfused rat hindlimb muscles causes violent muscular contraction resulting in a stimulation of lactate production but a decrease in fructose-2,6-bisphosphate.[75] In more recent experiments carried out in vivo,[95] it was shown that electrical stimulation of gastrocnemius resulted in a transient twofold increase in fructose-2,6-bisphosphate when the frequency of stimulation was lower than 5/sec, whereas its concentration decreased when higher frequencies were applied. The stimulation of lactate production and glycogen breakdown was minimal in the first case but maximal in the latter. It therefore appears that in these models of muscular contraction, there is no relationship between the rate of glycolysis and the concentration of fructose-2,6-bisphosphate.[95]

On the other hand, when rat hindlimb muscles were perfused with adrenaline or insulin, the production of lactate and the concentration of fructose-2,6-bisphosphate were increased.

Fructose-2,6-bisphosphate could therefore play a role in the control of glycolysis under these conditions.[75]

An activation of phosphofructokinase has been reported to occur in heart and muscle after adrenaline treatment.[96,97] It is not known whether fructose-2,6-bisphosphate is involved in this process. Finally it should be added that the phosphorylation of muscle phosphofructokinase by cyclic AMP-dependent protein kinase causes little change in the kinetic properties of the enzyme.[98]

VI. ROLE IN ADIPOSE TISSUE

Glycolysis in adipose tissue can be stimulated by adrenaline and insulin.[99] Treatment of isolated adipocytes with noradrenaline caused modifications in the kinetic properties of phosphofructokinase: the affinity for fructose 6-phosphate was increased whereas the V_{max} was decreased.[100] These changes could not be explained in terms of fructose-2,6-bisphosphate the concentration of which was not changed by noradrenaline.[101] It was found that free fatty acids are very potent stimulators of the enzyme and it was concluded that free fatty acids rather than fructose-2,6-bisphosphate are the intermediary of the effect of noradrenaline.[101] However, it is not clear whether the increase in fatty acids caused by noradrenaline-stimulated lipolysis results in an increase in the concentration of free acids or whether the bulk fatty acids in the cells are bound to proteins.

Recent experiments also showed that insulin treatment of fat pads stimulated the glycolytic flux and changed the kinetic properties of phosphofructokinase: the enzyme became less susceptible to inhibition by ATP and the V_{max} increased.[102] Here again there was no relationship between the increase in glycolytic flux and the concentration of fructose-2,6-bisphosphate.

It therefore appears that in white adipose tissue, fructose-2,6-bisphosphate cannot be considered as an important mediator of noradrenaline or insulin action on glycolysis.

VII. METABOLIC DETOURS FROM THE GLYCOLYTIC PATHWAY

There is some similarity between fructose-2,6-bisphosphate and some other diphosphoesters related to the glycolytic pathway. Like fructose-2,6-bisphosphate, glucose-1,6-bisphosphate and 2,3-bisphosphoglycerate are not true intermediates of the glycolysis and their synthesis and degradation are detours from the glycolytic pathway. Neither are they direct biosynthetic precursors of the pathway. Except for 2,3-bisphosphoglycerate in erythrocytes, their tissue concentration is in the μM range.[75,85,103,104] They are synthesized from a glycolytic intermediate which is either a substrate or a product of the reaction that they control. They act mainly as cosubstrates or effectors of reactions in the pathway. Although effects have been described on other systems,[103,105] it, however, seems that 2,3-bisphosphoglycerate is the only one which displays an important metabolic effect outside glycolysis: it modulates the affinity of hemoglobin for oxygen.[106] Among these diphosphoesters, fructose-2,6-bisphosphate is the only one for which no other effect has yet been well demonstrated.

The comparison of fructose-2,6-bisphosphate with 2,3-bisphosphoglycerate can be extended to the reactions responsible for their synthesis and degradation. The synthesis and degradation of 2,3-bisphosphoglycerate can be achieved by a multifunctional enzyme which displays phosphoglycerate mutase, bisphosphoglycerate synthase, and phosphatase activities.[107] The three enzymatic activities appear to be catalyzed at a common active site involving a histidyl residue which is phosphorylated by 2,3-bisphosphoglycerate or 1,3-bisphosphoglycerate. The phosphoryl-enzyme can (1) exchange the phosphoryl group with glycerate 3-phosphate or glycerate 2-phosphate regenerating the phosphoryl enzyme; (2) transfer the phosphoryl group to glycerate 3-phosphate, liberating the free enzyme and 2,3-bisphosphoglycerate; and (3) transfer the phosphoryl group to water. These steps can be summarized as follows:

2,3-bisphosphoglycerate + E $\rightleftharpoons$ E-P $\leftarrow$ 1,3-bisphosphoglycerate + E

E-P + glycerate 3-phosphate $\rightarrow$ E-P + glycerate 2-phosphate
or glycerate 2-phosphate or glycerate 3-phosphate (1)

E-P + glycerate 3-phosphate $\rightarrow$ E + 2,3-bisphosphoglycerate (2)

E-P + H_2O $\rightarrow$ E + P_i (3)

The synthesis and degradation of fructose-2,6-bisphosphate in the liver is catalyzed by two enzymes which have not been separated from each other, even after purification to homogeneity. This protein is now considered to be a multifunctional enzyme and referred to as phosphofructokinase-2/fructose-2,6-bisphosphatase.[43,108] The mechanism of the reaction is not yet fully understood but it appears that there is a histidyl residue which can be phosphorylated either by fructose-2,6-bisphosphate or ATP plus fructose 6-phosphate.[43] Since it has not been unequivocally shown that ATP alone can phosphorylate the enzyme, the existence of an unique active site remains to be demonstrated. The ATPase activity measured by Pilkis and co-workers[108] could then result from a combination of two activities, one catalyzing the synthesis of fructose-2,6-bisphosphate and the other its hydrolysis.

The multifunctional enzyme responsible for the synthesis and degradation of these two diphosphoesters share some common properties:

1. They have different activities on the same protein.
2. A phosphoryl-enzyme intermediate on a histidyl residue is formed during the reaction. However, there is no clear demonstration that the two activities of phosphofructokinase-2/fructose-2,6-bisphosphatase are catalyzed at a common single active site.
3. The substrates of one reaction are inhibitors of the other.
4. There are specific stimulators of the phosphatase reaction (2,3-bisphosphoglycerate phosphatase: glycolate 2-phosphate;[109] fructose-2,6-bisphosphatase: sn-glycerol 3-phosphate[37]) which probably bind to the substrate sites and thus prevent the phosphoryl transfer.

We tend to believe that the striking analogy existing between fructose-2,6-bisphosphate and 2,3-bisphosphoglycerate is not fortuitous. It may indicate that metabolic detours are a more general mechanism developed by the cell to synthesize effectors which control a metabolic pathway.

VIII. CONCLUSIONS

Fructose-2,6-bisphosphate is a very powerful stimulator of phosphofructokinase and an inhibitor of fructose-1,6-bisphosphatase. Its role in the control of glycolysis is well documented in liver where it switches gluconeogenesis to glycolysis when glucose is in excess. It is however not involved in the stimulation of glycolysis by anoxia, during tetanic contraction or after noradrenaline or insulin treatment of adipose tissue, but may play a role in the stimulation of glycolysis in muscle treated with insulin or adrenaline. It therefore appears that despite its promising properties, fructose-2,6-bisphosphate has been misleading and cannot be considered as the ubiquitous regulator of glycolysis.

ACKNOWLEDGMENTS

We are grateful to Drs. Y. Minatogawa and M. H. Rider for reading this manuscript and to Mrs. Th. Lambert and K. Wauters for their skillful typing of the manuscript. L. H. is Maître de Recherches du Fonds National de la Recherche Scientifique and R. B. was recipient of a Long-Term European Molecular Biology Organization Fellowship.

REFERENCES

1. **Van Schaftingen, E., Hue, L., and Hers, H. G.,** Fructose 2,6-bisphosphate, the probable structure of the glucose- and glucagon-sensitive stimulator of phosphofructokinase, *Biochem. J.*, 192, 897, 1980.
2. **Castaño, J. C., Nieto, A., and Felíu, J. E.,** Inactivation of phosphofructokinase by glucagon in rat hepatocytes, *J. Biol. Chem.*, 254, 5576, 1979.
3. **Kagimoto, T. and Uyeda, K.,** Hormone-stimulated phosphorylation of liver phosphofructokinase in vivo, *J. Biol. Chem.*, 254, 5584, 1979.
4. **Furuya, E. and Uyeda, K.,** An activation factor of liver phosphofructokinase, *Proc. Natl. Acad. Sci. U.S.A.*, 77, 5861, 1980.
5. **Claus, T. H., Schlumpf, J. R., El-Maghrabi, M. R., Pilkis, J., and Pilkis, S. J.,** Mechanism of action of glucagon on hepatocyte phosphofructokinase activity, *Proc. Natl. Acad. Sci. U.S.A.*, 77, 6501, 1980.
6. **Van Schaftingen, E., Hue, L., and Hers, H. G.,** Control of fructose 6-phosphate/fructose 1,6-bisphosphate cycle in isolated hepatocytes by glucose and glucagon, *Biochem. J.*, 192, 887, 1980.
7. **Van Schaftingen, E. and Hers, H. G.,** Synthesis of a stimulator of phosphofructokinase, most likely fructose 2,6-bisphosphate, from phosphoric acid and fructose 6-phosphoric acid, *Biochem. Biophys. Res. Commun.*, 96, 1524, 1980.
8. **Van Schaftingen, E. and Hers, H. G.,** Formation of fructose 2,6-bisphosphate from fructose 1,6-bisphosphate by intramolecular cyclisation followed by alkaline hydrolysis, *Eur. J. Biochem.*, 117, 319, 1981.
9. **Pilkis, S. J., El-Maghrabi, M. R., Pilkis, J., Claus, T. H., and Cumming, D.,** Fructose 2,6-bisphosphate, a new activator of phosphofructokinase, *J. Biol. Chem.*, 256, 3171, 1981.
10. **Uyeda, K., Furuya, E., and Sherry, A. D.,** The structure of "activation factor" for phosphofructokinase, *J. Biol. Chem.*, 256, 8679, 1981.
11. **Hesbain-Frisque, A., Van Schaftingen, E., and Hers, H. G.,** Structure and configuration of fructose 2,6-bisphosphate by ^{31}P and ^{13}C nuclear magnetic resonance, *Eur. J. Biochem.*, 117, 325, 1981.
12. **Hers, H. G. and Van Schaftingen, E.,** Fructose 2,6-bisphosphate 2 years after its discovery, *Biochem. J.*, 206, 1, 1982.
13. **Pilkis, S. J., El-Maghrabi, M. R., McGrane, M., Pilkis, J., Fox, E., and Claus, T. H.,** Fructose 2,6-bisphosphate: a mediator of hormone action at the fructose 6-phosphate/fructose 1,6-bisphosphate substrate cycle, *Mol. Cell Endocrinol.*, 25, 245, 1982.
14. **Uyeda, K., Furuya, E., Richard, C. S., and Yokoyama, M.,** Fructose-2,6-P_2, chemistry and biological function, *Mol. Cell. Biochem.*, 48, 97, 1982.
15. **Hers, H. G., Hue, L., and Van Schaftingen, E.,** Fructose 2,6-bisphosphate, *TIBS*, 9, 329, 1982.
16. **Hers, H. G. and Hue, L.,** Gluconeogenesis and related aspects of glycolysis, *Annu. Rev. Biochem.*, 52, 617, 1983.
17. **Van Schaftingen, E., Jett, M. F., Hue, L., and Hers, H. G.,** Control of liver 6 phosphofructokinase by fructose 2,6-bisphosphate and other effectors, *Proc. Natl. Acad. Sci. U.S.A.*, 78, 3483, 1981.
18. **Uyeda, K., Furuya, E., and Luby, L. J.,** The effect of natural and synthetic D-fructose 2,6-bisphosphate on the regulatory kinetic properties of liver and muscle phosphofructokinases, *J. Biol. Chem.*, 256, 8394, 1981.
19. **Aragon, J. J., Tornheim, K., and Lowenstein, J. M.,** On a possible role of IMP in the regulation of phosphorylase activity in skeletal muscle, *FEBS Lett.*, 117 (Suppl.), K56, 1980.
20. **Pilkis, S. J., El-Maghrabi, M. R., Pilkis, J., and Claus, T.,** Inhibition of fructose-1,6-bisphosphatase by fructose 2,6-bisphosphate, *J. Biol. Chem.*, 256, 3619, 1981.
21. **Van Schaftingen, E. and Hers, H. G.,** Inhibition of fructose 1,6–bisphosphatase by fructose 2,6–bisphosphate, *Proc. Natl. Acad. Sci. U.S.A.*, 78, 2861, 1981.
22. **Pilkis, S. J., El-Maghrabi, M. R., McGrane, M. M., Pilkis, J., and Claus, T. H.,** The role of fructose 2,6-bisphosphate in regulation of fructose-1,6-bisphosphatase, *J. Biol. Chem.*, 256, 11489, 1981.

23. **Gottschalk, M. E., Chatterjee, T., Edelstein, J., and Marcus, F.,** Studies on the mechanism of interaction of fructose 2,6-bisphosphate with fructose-1,6-bisphosphatase, *J. Biol. Chem.*, 257, 8016, 1982.
24. **François, J., Van Schaftingen, E., and Hers, H. G.,** On the mechanism of inhibition of neutral liver fructose 1,6-bisphosphatase by fructose 2,6-bisphosphate, *Eur. J. Biochem.*, 134, 269, 1983.
25. **Kitajima, S. and Uyeda, K.,** A binding study of the interaction of β-D-fructose 2,6-bisphosphate with phosphofructokinase and fructose-1,6-bisphosphatase, *J. Biol. Chem.*, 258, 7352, 1983.
26. **Van Schaftingen, E., Hue, L., and Hers, H. G.,** Study of the fructose 6-phosphate/fructose 1,6-bisphosphate cycle in the liver in vivo, *Biochem. J.*, 192, 263, 1980.
27. **Akerboom, T. P. M., Bookelman, H., Zuurendonk, P. F., Van Der Meer, R., and Tager, J. M.,** Intramitochondrial and extramitochondrial concentrations of adenine nucleotides and inorganic phosphate in isolated hepatocytes from fasted rats, *Eur. J. Biochem.*, 84, 413, 1978.
28. **Sols, A. and Marco, R.,** Concentrations of metabolites and binding sites. Implications in metabolic regulation, *Curr. Top. Cell. Regul.*, 2, 227, 1970.
29. **Sabularse, D. C. and Anderson, R. L.** D-fructose 2,6-bisphosphate: a naturally occurring activator for inorganic pyrophosphate: D-fructose-6-phosphate 1-phosphotransferase in plants, *Biochem. Biophys. Res. Commun.*, 103, 848, 1981.
30. **Van Schaftingen, E., Lederer, B., Bartrons, R., and Hers, H. G.,** A kinetic study of pyrophosphate: fructose 6-phosphatephosphotransferase from potato tubers, *Eur. J. Biochem.*, 129, 191, 1982.
31. **Gibson, D. M. and Shine, W. E.,** Uridine diphosphate glucose breakdown is mediated by a unique enzyme activated by fructose 2,6-bisphosphate in Solanum tuberosum, *Proc. Natl. Acad. Sci. U.S.A.*, 80, 2491, 1983.
32. **Van Schaftingen, E. and Hers, H. G.,** Phosphofructokinase. II. The enzyme that forms fructose 2,6-bisphosphate from fructose 6-phosphate and ATP, *Biochem. Biophys. Res. Commun.*, 101, 1078, 1981.
33. **Furuya, E. and Uyeda, K.,** A novel enzyme catalyzes the synthesis of activation factor from ATP and D-fructose-6-P, *J. Biol. Chem.*, 256, 7109, 1981.
34. **El-Maghrabi, M. R., Claus, T. H., Pilkis, J., and Pilkis, S. J.,** Partial purification of a rat liver enzyme that catalyzes the formation of fructose 2,6-bisphosphate, *Biochem. Biophys. Res. Commun.*, 101, 1071, 1981.
35. **Hue, L., Blackmore, P. F., and Exton, J. H.,** Fructose 2,6-bisphosphate: hormonal regulation and mechanism of its formation in liver, *J. Biol. Chem.*, 256, 8900, 1981.
36. **Furuya, E., Yokoyama, M., and Uyeda, K.,** An enzyme that catalyses hydrolysis of fructose 2,6-bisphosphate, *Biochem. Biophys. Res. Commun.*, 105, 264, 1982.
37. **Van Schaftingen, E., Davies, D. R., and Hers, H. G.,** Fructose 2,6-bisphosphatase from rat liver, *Eur. J. Biochem.*, 124, 143, 1982.
38. **El-Maghrabi, M. R., Claus, T. H., Pilkis, J., Fox, E., and Pilkis, S. J.,** Regulation of rat liver fructose 2,6-bisphosphatase, *J. Biol. Chem.*, 257, 7603, 1982.
39. **Van Schaftingen, E., Davies, D. R., and Hers, H. G.,** Inactivation of phosphofructokinase 2 by cyclic AMP-dependent protein kinase, *Biochem. Biophys. Res. Commun.*, 103, 362, 1981.
40. **El-Maghrabi, M. R., Fox, E., Pilkis, J., and Pilkis, S. J.,** Cyclic AMP-dependent phosphorylation of rat liver 6-phosphofructo 2-kinase/fructose 2,6-bisphosphatase, *Biochem. Biophys. Res. Commun.*, 106, 794, 1982.
41. **El-Maghrabi, M. R., Claus, T. H., Pilkis, J., and Pilkis, S. J.,** Regulation of 6-phosphofructo-2-kinase activity by cyclic AMP-dependent phosphorylation, *Proc. Natl. Acad. Sci. U.S.A.*, 79, 315, 1982.
42. **Furuya, E., Yokoyama, M., and Uyeda, K.,** Regulation of fructose-6-phosphate 2-kinase by phosphorylation and dephosphorylation: possible mechanism for coordinated control of glycolysis and glycogenolysis, *Proc. Natl. Acad. Sci. U.S.A.*, 79, 325, 1982.
43. **Pilkis, S. J., Walderhaug, M., Murray, K., Beth, A., Venkataramu, S. D., Pilkis, J., and El-Maghrabi, M. R.,** 6-Phosphofructo-2-kinase/fructose 2,6-bisphosphatase from rat liver, *J. Biol. Chem.*, 258, 6135, 1983.
44. **Bartrons, R., Hue, L., Van Schaftingen, E., and Hers, H. G.,** Hormonal control of fructose 2,6-bisphosphate concentration in isolated rat hepatocytes, *Biochem. J.*, 214, 829, 1983.
45. **Richards, C. S., Furuya, E., and Uyeda, K.,** Regulation of fructose 2,6-P_2 concentration in isolated hepatocytes, *Biochem. Biophys. Res. Commun.*, 100, 1673, 1981.
46. **Richards, C. S., Yokoyama, M., Furuya, E., and Uyeda, K.,** Reciprocal changes in fructose-6-phosphate, 2-kinase and fructose-2,6-bisphosphatase activity in response to glucagon and epinephrine, *Biochem. Biophys. Res. Commun.*, 104, 1073, 1982.
47. **Richards, C. S. and Uyeda, K.,** Hormonal regulation of fructose-6-P, 2-kinase and fructose-2,6-P_2 by two mechanisms, *J. Biol. Chem.*, 257, 8854, 1981.
48. **Engström, L.,** The regulation of liver pyruvate kinase by phosphorylation-dephosphorylation, *Curr. Top. Cell. Regul.*, 13, 29, 1978.

49. **Hue, L. and Bartrons, R.**, Role of fructose 2,6-bisphosphate in the control by glucagon of gluconeogenesis from various precursors in isolated hepatocytes, *Biochem. J.*, 218, 165, 1984.
50. **Exton, J. H., Mallette, L. E., Jefferson, L. S., Wong, E. H. A., Friedmann, N., Miller, T. B., and Park, C. R.**, The hormonal control of hepatic gluconeogenesis, *Rec. Prog. Horm. Res.*, 26, 411, 1970.
51. **Blair, J. B., Cimbala, M. A., Foster, J. L., and Morgan, R. A.**, Hepatic pyruvate kinase, *J. Biol. Chem.*, 251, 3756, 1976.
52. **Felíu, J. E., Hue, L., and Hers, H. G.**, Hormonal control of pyruvate kinase activity and of gluconeogenesis in isolated hepatocytes, *Proc. Natl. Acad. Sci. U.S.A.*, 73, 2762, 1976.
53. **Pilkis, S. J., Chrisman, T. D., El-Maghrabi, M. R., Colosia, A., Fox, E., Pilkis, J., and Claus, T. H.**, The action of insulin on hepatic fructose 2,6-bisphosphate metabolism, *J. Biol. Chem.*, 258, 1495, 1982.
54. **Bartrons, R., Van Schaftingen, E., and Hers, H. G.**, The property of adenosine to decrease the concentration of fructose 2,6-bisphosphate in isolated hepatocytes. A cyclic AMP-mediated effect, *Biochem. J.*, 218, 157, 1984.
55. **Hue, L., Van Schaftingen, E., and Blackmore, P. F.**, Stimulation of glycolysis and accumulation of a stimulator of phosphofructokinase in hepatocytes incubated with vasopressin, *Biochem. J.*, 194, 1023, 1981.
56. **Exton, J. H.**, Molecular mechanisms involved in alpha-adrenergic responses, *Mol. Cell. Endocrinol.*, 23, 233, 1981.
57. **Williamson, J. R., Cooper, R. H., and Hoek, J. B.**, Role of calcium in the hormonal regulation of liver metabolism, *Biochim. Biophys. Acta*, 639, 243, 1981.
58. **Hue, L., Felíu, J. E., and Hers, H. G.**, Control of gluconeogenesis and of enzymes of glycogen metabolism in isolated rat hepatocytes, *Biochem. J.*, 176, 791, 1978.
59. **Hems, D. A., McCormack, J. G., and Denton, R. M.**, Activation of pyruvate dehydrogenase in the perfused rat liver by vasopressin, *Biochem. J.*, 176, 627, 1978.
60. **Assimacopoulos-Jeannet, F., McCormack, J. G., and Jeanrenaud, B.**, Effect of phenylephrine on pyruvate dehydrogenase activity in rat hepatocytes and its interaction with insulin and glucagon, *FEBS Lett.*, 159, 83, 1983.
61. **Rognstad, R. and Katz, J.**, Role of pyruvate kinase in the regulation of gluconeogenesis from L-lactate, *J. Biol. Chem.*, 252, 1831, 1977.
62. **Blair, J. B., James, M. E., and Foster, J. L.**, Adrenergic control of glucose output and adenine 3'-5'-monophosphate levels in hepatocytes from juvenile and adult rats, *J. Biol. Chem.*, 254, 7579, 1979.
63. **Studer, R. K. and Borle, A. B.**, Differences between male and female rats in the regulation of hepatic glycogenolysis, *J. Biol. Chem.*, 257, 7957, 1982.
64. **Woods, H. F. and Krebs, H. A.**, Lactate production in the perfused rat liver, *Biochem. J.*, 125, 129, 1971.
65. **Brunengraber, H., Boutry, M., and Lowenstein, J. M.**, Fatty acid and 3-β-hydroxysterol synthesis in the perfused rat liver, *J. Biol. Chem.*, 248, 2656, 1973.
66. **Hers, H. G.**, The control of glycogen metabolism in the liver, *Annu. Rev. Biochem.*, 45, 167, 1976.
67. **Stalmans, W.**, The hepatic threshold to glucose, *Curr. Top. Cell. Regul.*, 11, 51, 1976.
68. **Hue, L.**, Futile cycles and regulation of metabolism, in *Metabolic Compartmentation*, Sies, H., Ed., Academic Press, London, 1982, 71.
69. **Krebs, H. A.**, The Pasteur effect and the relations between respiration and fermentation, *Essays Biochem.*, 8, 1, 1972.
70. **Ramaiah, A.**, Pasteur effect and phosphofructokinase, *Curr. Top. Cell. Regul.*, 8, 297, 1974.
71. **Sols, A.**, The Pasteur effect in the allosteric era, in *Reflections on Biochemistry*, Kornberg, A., Horecker, B. L., Cornudella, L., and Oro, J., Eds., Pergamon Press, Oxford, 1976, 199.
72. **Cori, C. F.**, The role of lactic acid in the development of biochemistry, in *Reflections on Biochemistry*, Kornberg, A., Horecker, B. L., Cornudella, L., and Oro, J., Eds., Pergamon Press, New York, 1976, 17.
73. **Newsholme, E. A. and Start, C.**, *Regulation in Metabolism*, John Wiley & Sons, London, 1973, chap. 3.
74. **Hue, L.**, Role of fructose 2,6-bisphosphate in the stimulation of glycolysis by anoxia in isolated hepatocytes, *Biochem. J.*, 206, 359, 1982.
75. **Hue, L., Blackmore, P. F., Shikama, H., Robinson-Steiner, A., and Exton, J. H.**, Regulation of fructose-2,6-bisphosphate content in rat hepatocytes, perfused hearts and perfused hindlimbs, *J. Biol. Chem.*, 257, 4308, 1982.
76. **Van den Berghe, G.**, Metabolic effects of fructose in the liver, *Curr. Top. Cell. Regul.*, 13, 97, 1978.
77. **Solanki, K., Nyfeler, F., Mozer, U. K., and Walter, P.**, Effect of glucose on carbohydrate synthesis from alanine or lactate in hepatocytes from starved rats, *Biochem. J.*, 192, 377, 1980.
78. **Rognstad, R.**, Control of lactate gluconeogenesis by glucose in rat hepatocytes, *Arch. Biochem. Biophys.*, 217, 498, 1982.

79. **Claus, H. T., Schlumpf, J. R., El-Maghrabi, M. R., and Pilkis, S. J.,** Regulation of the phosphorylation and activity of 6-phosphofructo 1-kinase in isolated hepatocytes by α-glycerolphosphate and fructose 2,6-bisphosphate, *J. Biol. Chem.*, 257, 7541, 1982.
80. **Krebs, H. A. and Eggleston, L. V.,** The role of pyruvate kinase in the regulation of gluconeogenesis, *Biochem. J.*, 94, 3C, 1965.
81. **Felíu, J. E., Hue, L., and Hers, H. G.,** Regulation in vitro and in vivo of adenosine 3':5'-monophosphate-dependent inactivation or rat-liver pyruvate kinase type L, *Eur. J. Biochem.*, 81, 609, 1977.
82. **Van Berkel, T. J. C., Kruijt, J. K., and Koster, J. F.,** Hormone-induced changes in pyruvate kinase. Effects of glucagon and starvation, *Eur. J. Biochem.*, 81, 423, 1977.
83. **Dunaway, G. A., Leung, G. L. Y., Thrasher, J. R., and Cooper, M. D.,** Turnover of hepatic phosphofructokinase in normal and diabetic rats, *J. Biol. Chem.*, 253, 7460, 1978.
84. **Neely, P., El-Maghrabi, M. R., Pilkis, S. J., and Claus, T. H.,** Effect of diabetes, insulin, starvation, and refeeding on the level of rat hepatic fructose 2,6-bisphosphate, *Diabetes*, 30, 1062, 1981.
85. **Kuwajima, M. and Uyeda, K.,** The tissue distribution of fructose 2,6-P_2 and fructose-6-P, 2 kinase in rats and the effects of starvation, diabetes and hypoglycemia on hepatic fructose 2,6-P_2 and fructose-6-P, 2 kinase, *Biochem. Biophys. Res. Commun.*, 104, 84, 1982.
86. **Van Schaftingen, E. and Hers, H. G.,** The role of fructose 2-6-bisphosphate in the long-term control of phosphofructokinase in rat liver, *Biochem. Biophys. Res. Commun.*, 113, 548, 1983.
87. **Dunaway, G. A. and Segal, H. L.,** A stabilizing factor for liver phosphofructokinase, *Biochem. Biophys. Res. Commun.*, 56, 689, 1974.
88. **Dunaway, G. A. and Segal, H. L.,** Purification and physiological role of a peptide stabilizing factor of rat liver phosphofructokinase, *J. Biol. Chem.*, 251, 2323, 1976.
89. **Hue, L. and van de Werve, G.,** Increased concentration of fructose 2,6-bisphosphate in livers of genetically obese mice, *FEBS Lett.*, 145, 263, 1982.
90. **Claus, T. H. and Pilkis, S. J.,** Fructose 2,6-bisphosphate levels are elevated in livers of genetically obese mice, *Biochem. Biophys. Res. Commun.*, 109, 664, 1982.
91. **Hue, L., van de Werve, G., and Jeanrenaud, B.,** Fructose 2,6-bisphosphate in livers of genetically obese rats, *Biochem. J.*, 214, 1019, 1983.
92. **McCune, S. A., Durant, P. J., Jenkins, P. A., and Harris, R. A.,** Comparative studies on fatty acid synthesis, glycogen metabolism and gluconeogenesis by hepatocytes isolated from lean and obese Zucker rats, *Metabolism*, 30, 1170, 1981.
93. **Lavine, R. L., Voyles, N., Perrino, P. V., and Recant, L.,** The effect of fasting on tissue cyclic AMP and plasma glucagon in the obese hyperglycemic mouse, *Endocrinology*, 97, 615, 1975.
94. **Richter, E. A., Ruderman, N. B., Gavras, H., Belur, E. R., and Galbo, H.,** Muscle glycogenolysis during exercise: dual control by epinephrine and contractions, *Am. J. Physiol.*, 242, E25, 1982.
95. **Minatogawa, Y. and Hue, L.,** Fructose 2,6-bisphosphate in rat skeletal muscle during contraction, *Biochem. J.*, 223, 73, 1984.
96. **Clark, M. G. and Patten, G. S.,** Epinephrine activation of phosphofructokinase in perfused rat heart independent of changes in effector concentrations, *J. Biol. Chem.*, 256, 27, 1981.
97. **Mansour, T. E.,** Phosphofructokinase activity in skeletal muscle extracts following administration of epinephrine, *J. Biol. Chem.*, 247, 6059, 1972.
98. **Foe, L. G. and Kemp, G. K.,** Properties of phospho and dephospho forms of muscle phosphofructokinase, *J. Biol. Chem.*, 257, 6368, 1982.
99. **Saggerson, E. D. and Greenbaum, A. L.,** The regulation of triglyceride synthesis and fatty acid synthesis in rat epididymal adipose time, *Biochem. J.*, 119, 193, 1970.
100. **Soorana, S. R. and Saggerson, E. D.,** Rapid modulation of adipocyte phosphofructokinase activity by adrenalin and insulin, *Biochem. J.*, 202, 753, 1982.
101. **Lederer, B. and Hers, H. G.,** The mechanism by which nonepinephrine increases the activity of phosphofructo 1-kinase in isolated rat adipocytes, *FEBS Meeting*, S04/TH050, Brussels, 1983.
102. **Rider, M. H. and Hue, L.,** Regulation of fructose 2,6-bisphosphate concentration in white adipose tissue, *Biochem. J.*, in press.
103. **Beitner, R.,** The role of glucose 1,6-bisphosphate in the regulation of carbohydrate metabolism in muscle, *TIBS*, 4, 228, 1979.
104. **Sasaki, R., Ikura, K., Sugimotu, E., and Chiba, H.,** The microdetermination of 2,3-diphosphoglycerate, *Anal. Biochem.*, 61, 43, 1974.
105. **Chiba, H. and Sasaki, R.,** Functions of 2,3-bisphosphoglycerate and its metabolism, *Curr. Top. Cell. Regul.*, 14, 75, 1974.
106. **Sasaki, R., Ikura, K., Narita, H., Yanagawa, S., and Chiba, H.,** 2,3-bisphosphoglycerate in erythroid cells, *TIBS*, 7, 140, 1982.
107. **Rose, Z. B.,** The enzymology of 2,3-bisphosphoglycerate, *Adv. Enzymol.*, 51, 211, 1980.

108. **Pilkis, S. J., Chrisman, T., Burgress, B., McGrane, M., Colosia, A., Pilkis, J., Claus, T. H., and El-Maghrabi, M. R.,** Rat hepatic 6-phosphofructo 2-kinase/fructose 2,6-bisphosphatase: a unique bifunctional enzyme, *Adv. Enzymol. Regul.*, 21, 147, 1983.
109. **Rose, Z. B. and Liebowitz, J.,** 2,3-diphosphoglycerate phosphatase from human erythrocytes, *J. Biol. Chem.*, 245, 3232, 1970.

Chapter 3

REGULATION OF MAMMALIAN HEXOKINASE ACTIVITY

John E. Wilson

TABLE OF CONTENTS

I. INTRODUCTION

Hexokinase (ATP: D-hexose 6-phosphotransferase, EC 2.7.1.1) ushers glucose into the theater of cellular metabolism:

$$\text{Glucose} + \text{ATP} \cdot \text{Mg}^{++} \rightleftharpoons \text{Glucose-6-P} + \text{ADP} \cdot \text{Mg}^{++}$$

It is frequently the case that rather sophisticated regulatory influences are exerted at early steps in a metabolic sequence, and hexokinase does not disappoint us in this expectation. Although it is appropriate to consider transport into the cell as the initial step in glucose utilization, the relative importance of transport in regulation of glucose metabolism appears to depend upon the particular tissue.[1] In contrast, the properties of hexokinase are such as to suggest that this enzyme may exert a prominent regulatory influence on glucose metabolism in all mammalian tissues. It is the purpose of this chapter to review the properties of mammalian hexokinases and relate those properties to regulation of enzymatic activity in vivo.

A number of anabolic and catabolic pathways diverge from glucose-6-P, the product of the hexokinase reaction, and thus hexokinase can be considered to catalyze the first reaction in a variety of metabolic processes utilizing blood-borne glucose as substrate. Sols[2] has described glucose-6-P as a "crossroad" of cellular metabolism. Despite its involvement in several metabolic pathways, hexokinase is generally classified as a "glycolytic" enzyme, reflecting the role of this enzyme in a pathway of major importance for many mammalian tissues. Although such a classification is a convenient means for indicating a prominent metabolic role played by hexokinase, the involvement of this enzyme in other metabolic processes should not be disregarded. If glucose-6-P, rather than glucose, is viewed as the point of departure for various metabolic pathways, the known regulation of these alternative pathways becomes consistent with the commonly observed pattern mentioned above, i.e., that major regulatory influences are exerted on the initial steps of a pathway. Thus phosphofructokinase, generally accepted as of primary importance in regulation of glycolytic metabolism,[2,3] can then be appropriately viewed as the first "committed" reaction of glycolysis.[2] Similarly, if glucose-6-P rather than glucose is considered the substrate for the hexose monophosphate pathway, then regulation at the first reaction of the hexose monophosphate pathway, glucose-6-P dehydrogenase,[4,5] is also in accord with the commonly observed pattern of metabolic regulation.

Recognition of the fact that hexokinase catalyzes what is effectively the initial reaction in a number of alternative metabolic pathways confronts us with what could be a regulatory dilemma, for it is not apparent that all of these pathways would be regulated in a parallel manner. For example, increased glycolysis may be an appropriate response to increased energy demands resulting from cellular functions (e.g., ion movements, mobility) but such conditions are commonly associated with decreased biosynthetic activities which may be supplied with reducing equivalents by operation of the hexose monophosphate pathway, i.e., a redirection of the available metabolic energy from biosynthetic to nonbiosynthetic functions. It could, of course, be argued that increased glycolysis and decreased metabolism via the hexose monophosphate pathway are readily accomplished by simple diversion of the glucose-6-P produced by hexokinase into one pathway or the other by regulatory influences (activation of phosphofructokinase, inhibition of glucose-6-P dehydrogenase) acting after the glucose-6-P crossroad, i.e., additional regulation at the hexokinase step is not necessary. Implicit in this view is the assumption that glucose-6-P produced by hexokinase is equally available to the enzymes initiating the alternative metabolic fates. This assumption conflicts with increasing evidence for compartmentation of both enzymes and metabolites within the cell.[6-9] Thus it is not unreasonable to consider the possibility that distinct "pools" of hexokinase, having different intracellular locations and/or altered responses to metabolites influencing catalytic activity, may be associated with different metabolic functions (e.g., glycolysis vs. hexose monophosphate pathway).

It is not our intention to exhaustively review the extensive literature on mammalian hexokinases. For additional information, the reader is referred to earlier reviews by Colowick[10] and Purich et al.[11] Additional reviews dealing specifically with the hexokinase of brain are also available,[12,13] while Bessman and Geiger[14] have given particular attention to the possible regulatory significance of compartmentation of hexokinase.

II. ISOZYMES

A. Discovery and Nomenclature

There are, in fact, at least four distinct proteins capable of catalyzing the hexokinase reaction in mammalian tissues. Early reports of distinct glucose phosphorylating activities in rat liver,[15,16] differing markedly in their apparent affinity (K_m) for glucose, were quickly followed by demonstration of four forms of the enzyme distinguishable by chromatographic[17] or electrophoretic[18] methods. Further studies disclosed that similar multiplicity was present in other tissues, and also that the relative proportion of the various isozymes could be influenced by development or by fasting.[19]

There are two systems of nomenclature currently used to designate the multiple forms of hexokinase. González et al.[17] designated the four chromatographically distinguishable forms from rat liver as types A to D, based on their order of elution from a DEAE-cellulose column; this nomenclature system continues to be used by the Chilean group.[20] Katzen and Schimke[19] designated the multiple forms as types I to IV in order of their increasing mobility (toward the anode at pH 8.4) during starch gel electrophoresis. Since both elution from DEAE-cellulose and electrophoretic mobility can be related to charge on the molecule, it is not unexpected that types A to D of González et al.[17] correspond to types I to IV of Katzen and Schimke,[19] i.e., A = I, B = II, etc. Both systems of nomenclature were defined prior to the recommendations by the International Union of Biochemistry[21] and hence are,not surprisingly, in conflict with that recommendation which was to numerically designate multiple forms in order of *decreasing* mobility towards the anode. The designations defined by Katzen and Schimke[19] have been most widely utilized in the literature and we will adhere to that system in the present writing.

It may also be noted here that, in addition to types I to IV, the occurrence of a specific "sperm type" hexokinase has also been reported,[22] but there apparently has been little further characterization of this isozyme.

B. "High K_m" and "Low K_m": Glucokinase and Hexokinase

Although, as will be discussed in detail below, there are extensive similarities between the isozymes I to III (A to C in the nomenclature of González et al.[17]), there are notable differences between these isozymes and type IV. The most striking difference observed in the early studies[15-19] was in the K_m for glucose. Types I to III had K_m values in the 10^{-4} to 10^{-6} *M* range, markedly lower than the $K_m \approx 10^{-2}$ *M* exhibited by type IV. These observations gave rise to the distinction between "low K_m" (types I to III) and "high K_m" (type IV) hexokinases. More commonly, the latter enzyme is now referred to as "glucokinase" while the term "hexokinase" is usually taken to mean the "low K_m" forms;[23] we will subscribe to this practice in the present chapter. Despite the large difference in K_m values for glucose, all four isozymes exhibit similar K_m values (about 10^{-3} to 10^{-4} *M*) for the other substrate, ATP·Mg^{++}.[16,17,24]

Early studies indicated that glucokinase was found only in liver,[19,23] though this remains a controversial point.[25,26] What does seem to be generally accepted is that, in liver, the glucokinase is largely, and perhaps exclusively, located in the parenchymal cells, while the hexokinase activity is associated with nonparenchymal cells.[23,27-29] Meglasson et al.[26] have also found a glucokinase activity in pancreatic islets. Though some kinetic similarities with the liver glucokinase[30-32] were reported,[26] it remains to be shown whether or not the two enzymes are, in fact, the same gene product. Glucose utilization by pancreatic islets[33] and insulin secretion by the β-cells[34a] show similar sigmoidal responses to increasing glucose concentration, suggesting that the glucokinase, presumably responsible for the sigmoidal increase in glucose utilization,[30-33] was located in the insulin secreting β-cells. It seems likely that the hexokinase, also found in the islets in amounts (in terms of catalytic activity) comparable to the glucokinase,[26] may be localized in the other (i.e., not β-cells) cell types found in the islets; if this glucose phosphorylating activity coexisted with the glucokinase in the β-cells the sigmoidal dependence of glucose utilization[33] (dependent on phosphorylation) and insulin secretion[34a] on glucose concentration would likely have been disrupted. While definitive proof has yet to be obtained, the available results suggest that, with the possible exception of certain tumors,[34b,c] glucokinase and hexokinase may not coexist in any mammalian cell-type. The consequences of this are worth noting, for it implies that, in cells possessing only glucokinase (as is apparently the case for the hepatocyte, and possibly the β-cell), glucose metabolism and any physiological consequences dependent on that metabolism would be highly responsive to changes in blood glucose concentrations within the physiologically relevant range.

In addition to the marked difference in their K_m for glucose and tissue distribution, noted above, the hexokinase and glucokinase activities of liver present quite different developmental profiles. The liver of fetal or newborn rats has very low levels of glucokinase which increase markedly during development[16,35,36] whereas hexokinase levels decrease.[36]

Developmental increases in glucokinase are advanced by sequential injection of hydrocortisone followed by glucose while this treatment has no effect on hexokinase activity.[36] Differences in response of these activities to fasting[23,37] or to insulin administration in alloxan-diabetic rats have also been observed.[23] Coupled with other differences in their properties, discussed below, there is ample reason to regard the low K_m hexokinases and glucokinase as distinct species. Although, for comparative purposes, we will have occasion to mention glucokinase in other contexts, the principle focus for the remainder of this chapter will be the low K_m hexokinases, types I to III.

C. Further Complexity: Microheterogeneity of the Principal Isozymes of Hexokinase

There is a report of genetically determined polymorphism of the type III isozyme from human leukocytes.[38] This is apparently the only reported case of a genetically based heterogeneity in an isozyme of hexokinase from human tissues. Genetically determined polymorphism of an unspecified (but probably type I) isozyme in crab-eating macaques has also been reported.[39] There appear to have been no comparable reports of genetically based heterogeneity in the hexokinase isozymes from other animal species. Thus, in general, genetically determined polymorphism of the hexokinase isozymes is rather rare.

There are, however, several reports indicating microheterogeneity of the hexokinase isozymes, detectable by chromatographic or electrophoretic methods,[22,38,40-51] which is highly unlikely to be, and in some cases clearly is not, genetically based. This seems to be particularly common with the type I isozyme, with microheterogeneity having been observed with the enzyme from rat liver [41] and brain,[42,43] pig heart,[44] human[38,46,48] and rabbit[45,47] red blood cells, as well as other human tissues.[40,49] In the case of the human erythrocyte enzyme, a rather selective loss in one of the subforms (type I_b) is reported to occur as the cell ages.[46,48] The fact that the methods demonstrating the heterogeneity depend on charge differences (e.g., electrophoretic mobility or isoelectric point) has given rise to the expected speculation that deamidation of glutamine or asparagine residues, phosphorylation, or removal of charged residues by limited proteolysis may be responsible but there is, in fact, no definitive evidence as to the exact nature of the modification leading to the observed heterogeneity. However, for the rat brain enzyme, there is clear evidence that the heterogeneity is artifactual, occurring after the tissue is homogenized and very likely proteolytic in nature.[42,43] This may turn out to be the case with the type I isozyme from other tissues also.

The heterogeneity reported with the type II isozyme[22,51,52] is more limited than that seen with the type I hexokinase. Specifically, two subforms (II_a and II_b) have been observed and it seems clear that this heterogeneity is somehow related to sulfhydryl groups in the enzyme, i.e., although a single type II band is seen when electrophoresis is conducted in the presence of 2-mercaptoethanol (added to protect sulfhydryl groups in the enzyme), two bands are observed when 2-mercaptoethanol is deleted. The explanation for this effect and the possible physiological significance of II_a and II_b remains unclear.

Investigations thus far have failed to disclose any consistent differences between the various subforms of the hexokinase isozymes that might suggest different physiological roles. Therefore, although it is prudent to be aware of the existence and possible consequences of microheterogeneity within the major isozyme types, there is presently little reason to treat these subforms as discrete species of physiological relevance, and, indeed the extent to which such microheterogeneity is artifactual requires further attention. We will, therefore, speak of the "type I" and "type II" isozymes as single species and will refer to particular subforms only in specific instances when such distinctions are appropriate. Microheterogeneity of the type III isozyme has not been reported.

III. COMPARISON OF HEXOKINASE ISOZYMES

Although some may view the occurrence of distinct isozymes as physiologically irrelevant, presumably a more widely accepted view would be that the maintenance of these discrete forms in mammalian tissues indicates that they are somehow tailored to fulfill specific, and different, physiological roles. It may therefore be useful to compare the properties of the mammalian hexokinase isozymes in an attempt to discover differences that suggest physiological consequences.

A. Tissue Distribution

The isozyme patterns observed in several tissues from various mammalian species are compared in Table 1. Only tissues for which comparisons can be made between two or more species are included in the table; additional data on the isozymes found in other rat[22,58] and guinea pig[59] tissues, as well as normal and leukemic human lymphocytes and lymphoblasts,[60] other human blood cell types,[61] and Ehrlich ascites tumor cells[62] have also been reported. Considering that the results in Table 1 were reported by several different laboratories, there is rather good agreement. Several of the apparent discrepancies involve the relative amount of the type II isozyme. This isozyme is relatively unstable[22,24,63,64] and highly responsive to nutritional or hormonal status,[19,22,52,63] and it may be that these factors have influenced the observed isozyme pattern.

In considering these results, it is appropriate to bear in mind that the isozyme composition of a tissue does *not* necessarily reflect the isozyme composition of every cell within that tissue. Obviously tissues may include multiple cell types having different metabolic characteristics and it is reasonable to expect that the isozyme composition may reflect that metabolic diversity. For example, as discussed above, although rat liver is reported to contain both low K_m hexokinase as well as glucokinase,[15-20,22-24] these enzymes are located in different cell populations.[27-29] Indeed, for this reason we have purposely deleted liver from the tissues considered in Table 1. Our objective is to relate isozyme composition to metabolic properties

Table 1
DISTRIBUTION OF HEXOKINASE ISOZYMES IN MAMMALIAN TISSUES[a]

		Isozyme content			
Tissue	Source	High	Intermediate	Low	Ref.
Brain	Rat	I		II	22, 53a
	Guinea pig	I		III	50
	Guinea pig	I		II	53a
	Mouse	I		II	53a
	Hamster	I		II	53a
	Rabbit	I		II	53a
	Rabbit	I			53b
	Monkey	I		II	53a
	Cow	I		II	53a
	Human	I		II	49
Erythrocytes	Rat	I		II	22
	Rabbit	I		II, III	45
	Human	II	I		40
	Human	I		III	48
Kidney	Rat	I	III	II	22
	Guinea pig	I			50
	Rabbit	I		II	53b
	Human	I			49, 54
Lung	Rat	I, II, III			22
	Rat	I	II, III		55
	Rat	I	II	III	56
	Guinea pig	I, III	II		50
	Human	I	III		49, 54
Spleen	Rat	I, II, III			22
	Rabbit	I		II	53b
	Human	I, II			54
	Human	I	III		49
Heart	Rat	II	I	III	22
	Rabbit	I		II	53b
	Guinea pig	I	III	II	50
	Human	I			49, 54
Skeletal muscle	Rat	II		I, III	22
	Rat	II		I	53a, 57
	Mouse	II			57
	Bat	II		I	57
	Guinea pig	I		II, III	50
	Guinea pig	II	I		53a
	Guinea pig	II			57
	Rabbit	II	I		53a—b, 57
	Monkey	II	I		53a
	Cow	II		I	53a
	Sheep	II		I	57
	Human	I	II		40, 49
	Human	I			49

[a] As discussed in the text (Section III.C), an appreciable fraction of the type I isozyme and, to a lesser extent, the type II isozyme may be associated with the mitochondrial fraction from various tissues. However, virtually all studies on tissue isozyme composition (including all of those listed in this Table except for the work by Salotra and Singh[56] on rat lung) have used only the "soluble" fraction of tissue homogenates. Such a procedure may lead to underestimation of the relative proportion of the type I, and to a lesser extent, the type II isozyme, and an overestimation of the relative amount of the type III isozyme which seems to be restricted to the soluble fraction. This very likely accounts for the relatively greater amount of type III hexokinase reported for lung by investigators using high speed supernatants only,[22,49,50,54,55] whereas Salotra and Singh,[56] who considered both soluble and particulate fractions, found type III to be a minor component.

of the cells or tissues containing the isozymes. However, certainly most of the activities normally attributed to liver are associated with the hepatocytes, which contain little if any hexokinase.[27-29] It is obviously of questionable utility to attempt to relate hepatocyte metabolism to isozymic composition of nonhepatocytes. It might also be pointed out that, in contrast to other tissues which have been examined, the isozymes present in liver are highly species dependent,[20,57] suggesting no correlation between isozyme content (of the nonhepatocytes) and the principal metabolic functions of liver which seem to be similar in a variety of species.

Examination of Table 1 suggests certain correlations between isozyme content and the nature of tissue metabolism. First of all, it is evident that, with few exceptions, the type I isozyme is relatively high in all tissues except for skeletal muscle. This is also true for most other tissues or cells that have been examined.[22,59-62] It would seem, therefore, that the type I isozyme could be considered as the "basic" hexokinase, apparently having properties compatible with the differing metabolic characteristics associated with the diversity of sources examined. Indeed, it does not seem unreasonable to suppose that this isozyme plays an essentially identical role in all these tissues. In this regard, it is interesting to note the correlation between the relative predominance of the type I isozyme and the relative dependence on blood-borne glucose as a substrate for cellular metabolism. Thus, as is well known, brain is virtually totally dependent (under normal conditions) on blood-borne glucose as a substrate for its very high rate of energy metabolism, with virtually all of the glucose being metabolized via the glycolytic pathway,[12,13] and brain contains virtually exclusively the type I isozyme. Similarly, erythrocytes also are critically dependent on blood glucose as a substrate, metabolizing it predominantly via the glycolytic pathway under normal conditions[65] and they too contain, virtually exclusively, the type I isozyme. This isozyme also predominates in kidney and it has been shown that the relative levels of hexokinase in papillar, medullar, and cortical zones of the kidney are correlated with the relative rates of glucose utilization in these zones.[66] Thus we may infer that the essential physiological role of the type I isozyme is to provide glucose-6-P for a cellular metabolism, predominantly glycolytic in nature, in tissues (or cells) that are critically dependent on blood-borne glucose as substrate. We may expect an enzyme fulfilling such a role to have assured access to the second required substrate, $ATP \cdot Mg^{++}$, and to be highly responsive to altered energy status of the tissue. As we will see below, the type I isozyme is admirably suited for this role.

The observation that type I is the predominant isozyme in heart,[22,49,50,53b,54] a tissue which normally utilizes fatty acids as the preferred substrate for energy metabolism,[67,68] may seem to conflict with the above interpretation. However, it is well established that the heart can and does utilize glucose as an alternative substrate under partially ischemic or anoxic conditions.[67,69] Thus the constitutive presence of relatively high tissue levels of type I hexokinase[50,54,70] represents a capacity for glycolytic metabolism which can supplement the supply of metabolic energy obtainable from fatty acid oxidation (e.g., during periods of increased energy demand) and which may ensure continued provision of metabolic energy under temporary conditions (e.g., brief periods of limited oxygen availability) which hinder utilization of the normal preferred substrate, fatty acids. A somewhat analogous situation pertains in the developing brain where it has been shown that normal levels of hexokinase persist even under conditions in which ketone bodies supplant an appreciable portion of the normal glucose requirement.[71] It has been pointed out[71] that maintenance of this capacity for glucose utilization, even if in temporary disuse, would be critical to a tissue such as brain where the supply of the alternative substrate, ketone bodies, is highly susceptible to dietary status.

In contrast to the predominance of type I hexokinase in most other tissues examined, it is evident the type II is the major isozyme found in skeletal muscle (Table 1). In addition, type II hexokinase is the major form present in other insulin-sensitive tissues such as diaphragm,[22] mammary gland,[72] and adipose tissue.[73] Walters and McLean[72] reported that type II hexokinase activity of rat mammary gland changed in parallel with insulin sensitivity during the lactation cycle. Katzen[22] has drawn attention to this apparent relationship between the predominance of type II hexokinase and insulin sensitivity of the tissue, i.e., in highly insulin-sensitive tissues such as skeletal muscle, the majority of the glucose phosphorylating capacity is due to the type II isozyme; as the relative proportions of the type I and III (chiefly the type I) isozymes increase, the sensitivity to insulin decreases. Maintenance of tissue levels of type II hexokinase is highly dependent on insulin availability; marked decreases in type II hexokinase occur in insulin-sensitive tissues from diabetic animals.[22,52,55,63,72,73] Thus the ability to transport glucose, as governed by insulin in these tissues, is correlated with the capacity for phosphorylation of glucose, as reflected in type II hexokinase levels. In contrast to the constitutive nature of the type I isozyme, the type II isozyme represents a glucose phosphorylation capacity which is responsive to the availability of glucose, as determined by insulin-sensitive transport or nutritional status. As such, we may associate the type II isozyme with tissues (or cells) that do *not* consistently exhibit a dependence on blood-borne glucose as a substrate (or supplemental substrate) for energy metabolism. Type II hexokinase, therefore, represents an adaptation to the availability of glucose, which *may* be used for the generation of metabolic energy, or to support the synthesis of energy storage forms such as glycogen (skeletal muscle, diaphragm) or lipids (adipose tissues, mammary gland).

Unlike the type I and II isozymes, the type III hexokinase has not been found to be the predominant form in any tissue that has been examined (although this is not to say that it may not be the predominant form in certain cell types within a tissue). As with the type I but in contrast to the type II isozyme, type III hexokinase levels appear to be relatively unaffected by diabetes. Though a slight increase in type III activity was observed in heart of streptozotocin-diabetic rats, the type III activity in kidney, spleen, and liver was not altered.[52] McLean et al.[55] found type I and III activities as well as the type II isozyme to be unaltered in lung of alloxan-diabetic rats. Of the tissues that have been examined, only lung, and perhaps spleen, exhibit relatively high levels of the type III isozyme (Table 1). At present it is difficult to define any distinctive feature of glucose metabolism that might be relatively unique to lung and spleen, and for which the type III isozyme might be particularly well suited. Of the three isozymes, type III has the lowest K_m (about 10^{-6} *M*) for glucose and is also the only isozyme to exhibit marked substrate inhibition at glucose concentrations greater than about 0.3 m*M* glucose.[17,19,24] One might infer from this that the type III isozyme would be functional only at very low intracellular glucose levels. Additional studies on the cellular location of the type III isozyme as well as definition of the characteristics of glucose metabolism in those cells should lead to a better understanding of specific metabolic functions that may be associated with type III hexokinase.

B. Molecular Properties

All three isozymes of mammalian hexokinase, as well as glucokinase, have been purified to homogeneity from a number of tissues. The molecular weights and isoelectric points reported for these enzymes are summarized in Table 2; also included are molecular weights reported for other highly purified, but not homogeneous, preparations. All mammalian hexokinases that have been examined consist of a single polypeptide chain with a molecular weight of approximately 100,000.[47,76-83] This is approximately twice the molecular weight of glucokinase, which has given rise to speculation[20,44,82,84] that the mammalian hexokinases may represent the result of duplication and fusion of a gene for an ancestral glucokinase. A similar relationship between the mammalian hexokinases and the yeast or plant hexokinases which, like liver glucokinase, also consist of polypeptide chains of molecular weight about 50,000, had previously been suggested.[10,78] According to this view, the product of the original

Table 2
COMPARISON OF MOLECULAR WEIGHTS AND ISOELECTRIC POINTS REPORTED FOR HEXOKINASE ISOZYMES FROM VARIOUS MAMMALIAN TISSUES

Isozyme	Source	Homogeneous	Mol wt	Isoelectric point	Ref.
I	Bovine brain	Yes	ND[a]	ND	74
I	Bovine brain	Yes	107,000	ND	75
I	Rat brain	Yes	98,000	6.3	76
I	Rabbit brain	Yes	100,000	6.3	77
I	Porcine heart	Yes	97,000	~6.0	78, 79
I	Rat kidney	Yes	96,000	ND	80
I	Rabbit erythrocytes	Yes	110,000	6.2—6.3	81
I	Rabbit reticulocytes	Yes	105,000	5.7—5.8, 6.2—6.3 (subforms)	47
I	Human erythrocytes	No	100,000	5.35, 5.5, 5.7 (subforms)	48
I	Rat brain	No	96,000	ND	24
I	Rat kidney	No	99,000	ND	24
II	Rat skeletal muscle	Yes	96,000	ND	82
II	Rat skeletal muscle	Yes	106,000—108,000	5.7	83
II	Rat skeletal muscle	No	96,000	ND	24
II	Rat epididymal fat pad	No	90,000	ND	24
III	Rat liver	Yes	98,000	ND	80
III	Rat liver	No	99,500	ND	24
IV	Rat liver	Yes	48,000—52,000	ND	84
IV	Rat liver	Yes	53,000—57,000	ND	85
IV	Dog liver	Yes	47,000—49,000	ND	86

ND, not determined.

duplicated gene (i.e., the ancestor of present day mammalian hexokinase) would have had two catalytic sites, one of which could have evolved into an allosteric regulatory site for binding of glucose-6-P[44,78] (see below for a discussion of the allosteric site concept). Some support for such an evolutionary relationship can be found in Ureta's recent comparison[20] of the amino acid composition of mammalian hexokinases and glucokinase, using the theoretically rigorous comparative method developed by Cornish-Bowden.[87a] Whether or not the similarities are due to evolutionary relationships may be debatable, but Ureta's comparison nonetheless makes it clear that there are very great similarities in amino acid composition suggesting extensive homology between the mammalian hexokinases and glucokinase. Based on comparison of tyrosine-containing peptides by a peptide mapping technique, Darby and Trayer[87b] have reported some similarity between yeast hexokinase and the glucokinase and type II hexokinase from rat; this report was preliminary, however, and confirmation of the actual identity of these presumably common peptides will be necessary. Given these observations, it is not at all surprising that the amino acid compositions of the type I isozymes isolated from different mammalian tissues are also remarkably similar.[20,76,78]

The mammalian hexokinases are, based on their isoelectric points (Table 2), somewhat acidic in character, i.e., they would possess a net negative charge at neutral pH. The pI of the type II isozyme is lower than that for the type I isozyme, consistent with their relative electrophoretic mobility and order of elution from an anion-exchange column, the properties originally used to define the different types.[17-19] Although the isoelectric point of the type III isozyme has not been measured directly, its increased mobility[19] and later elution from DEAE-cellulose[17] indicate that it is still lower than that of the type II isozyme (pI = 5.7). Thus, at physiological pH, the molecular charges are expected to be negative and to increase in the order I < II < III. This may be related to the relative propensity of the isozymes for associating with membranous structures of the cell,[52] a subject which will be discussed in detail later in this chapter.

Despite the considerable similarities in amino acid composition,[20] suggesting corresponding similarities in amino acid sequence and structure,[87a] it is clear that glucokinase and the hexokinases as well as the hexokinase isozymes are immunologically quite distinct. Antibodies prepared against partially purified rat liver glucokinase also reacted with the glucokinase present in livers of a number of mammalian species (and even turtle liver glucokinase), but not with the type I, II, or III hexokinase activities in these same liver extracts.[20,88,89] Antibodies prepared against a partially purified type II isozyme from rat skeletal muscle did not inhibit the type I isozyme (from brain or adipose tissue) or liver glucokinase from this same species.[90] Some inhibition of hexokinase (presumably type I, see Table 1) in brain extracts of other species was seen with the anti-rat type II serum;[90] this appears to be the only study that has produced any results suggesting cross-reactivity between the different isozyme types from different species. In contrast, a number of investigators have reported that the isozymes are immunologically distinct whether compared in a single animal or across various species. Thus antisera against type I or type III isozymes of humans show no cross-reactivity, i.e., anti-type I does not react with the human type III isozyme, and vice versa.[49,54] Antibodies against purified type I hexokinase of bovine brain crossreact with the type I isozyme but not types II or III, or glucokinase from the rat.[91] Even greater selectivity has been reported, with Ureta and co-workers[20,92] observing that antibodies prepared against the purified type I isozyme of rat brain reacted poorly even with the type I isozyme of closely related rodents and not at all with the type I isozyme from cow, pig, or sheep; this anti-type I serum also did not cross-react with types II or III, or glucokinase from rat,[92] again confirming the immunological distinction of these various forms of glucose phosphorylating activity. As noted by Ureta,[20] the available results clearly suggest that glucokinases of various species bear highly conserved immunological determinants. In contrast, and despite possibly extensive homologies suggested by the striking similarities in amino acid composition, the type I, II, and III isozymes, and glucokinase are immunologically very distinct from each other. This immunological distinctiveness, coupled with a presumed extensive overall homology of the type I to III hexokinases, suggests that the major differences between the isozymes are found in surface features of the molecules since it is these features that commonly represent the immunological determinants[93,94] and determine the effective molecular charge and thus electrophoretic or chromatographic mobility which are notably different for these isozymes.[17-19]

C. Binding of Hexokinase to Intracellular Structures

The report of Crane and Sols,[95] which indicated the association of appreciable amounts of hexokinase with particulate elements in homogenates of brain and other tissues of the rat, has been confirmed and extended by several laboratories, and particulate forms of hexokinase have now been found in brain,[96-99] heart,[52,70,100,101] intestine,[70,102,103] lung,[56,70,104] epididymal fat pad,[105] adrenal medulla,[106] diaphragm,[52,70] skeletal muscle,[52,107] uterus,[108] mammary gland,[73,109] retina,[110] kidney[70] pancreas,[70] and spleen.[70] Although the subcellular structure with which hexokinase was associated was not determined in all of these studies (i.e., the "particulate" nature was not further defined), several investigators have shown that by far the major portion of the activity was associated with the mitochondria in most tissues examined.[56,70,96,97,99,101-107,109,110] More specific discussion of this binding to the mitochondria and its physiological consequences will be given later in this chapter. In the present context, we are concerned with possible differences between the isozymes in their ability to bind to mitochondria.

Katzen et al.[52] examined the hexokinase isozymes present in particulate fractions of rat heart, diaphragm, and skeletal muscle, and concluded that only the type I and II isozymes were associated with these fractions, any type III isozyme consistently being found only in the soluble fraction. In lung[56] and intestine[103] also, the type III isozyme was exclusively in the soluble fraction, while appreciable amounts of both type I and II hexokinase were associated with the mitochondria. In contrast to the apparent exclusively cytoplasmic location of the type III isozyme, a significant fraction of the total type I and II hexokinase activities were shown to be associated with mitochondria from, in addition to the tissues noted above, brain,[96,99] epididymal fat pad,[105] uterus,[108] and mammary gland.[109]

Kurokawa et al.[111] have recently shown that the type I and II isozymes of ascites tumor apparently compete for a common site on the mitochondrial membrane, i.e., saturation with one of the isozymes precludes additional binding of the other. These workers also took good advantage of a previous observation by Rose and Warms[97] that the ability of ascites tumor hexokinase to bind to mitochondria could be selectively destroyed by limited chymotryptic digestion while this treatment had *no* effect on the catalytic activity of the enzyme. Kurokawa et al.[111] found that chymotryptic treatment of both type I and II isozymes gave these same results, leading them to suggest that both isozymes might have a common (or very similar) chymotrypsin-sensitive structural feature which recognized a specific binding site on the mitochondria.

Based on an extensive study of the effects of neutral salts on the binding of type I hexokinase to mitochondria, Felgner and Wilson[112] concluded that there were two types of electrostatic interactions between the enzyme and the outer mitochondrial membrane. The first was a repulsive interaction between negatively charged groups on enzyme and membrane. These could be masked by divalent cations which could, in effect, provide a positively charged bridge between hexokinase and the membrane, thereby accounting for the enhancement of binding by divalent cations.[97,112,113] The second electrostatic interaction, attractive in nature, was between specific oppositely charged groups on enzyme and membrane, with disruption of these forces at high ionic strength accounting for the solubilization of the mitochondrial enzyme at high salt concentrations.[97,112-114] Based on these results,[112] increasing net negative charge on the hexokinase molecule would increase the repulsive interactions, thereby weakening or even preventing the binding of the enzyme. As noted in the above comparison of the pI values for the hexokinase isozymes, the molecular charge at neutral pH should become increasingly negative in the order I < II < III, and thus we would expect binding to *decrease* in this same order. This is consistent with available results. Thus, Kurokawa et al.[111] found that, in the absence of divalent cations (when repulsive interactions would be most effective), type I hexokinase bound to a much greater extent than did the Type II isozyme. Furthermore, as noted above, type III hexokinase has not been found to be associated with mitochondria in any tissue thus far examined. Binding is likely to be determined by surface features of the hexokinase molecule, which could be expected to interact most directly with the membrane surface, and, as discussed in Section III.B, it is precisely in this region of the molecule that the various isozymes are expected to be most different. It is likely that this relative ability to interact with the mitochondria, with its attendant metabolic consequences (discussed below), is a property which figures prominently in determining the isozyme pattern displayed by a particular tissue or cell type.

Hexokinase binds selectively to the outer mitochondrial membrane.[115-118] It has been suggested[12] that reports[119-121] of an appreciable amount of the mitochondrial hexokinase being associated with the inner mitochondrial membrane were in error due to technical difficulties. Nevertheless, the association of a small proportion of the mitochondrial activity with the inner mitochondrial membrane cannot be excluded, e.g., the recent report that about 10% of the mitochondrial hexokinase may be associated with the inner membrane of reticulocyte mitochondria.[122] An outer mitochondrial membrane binding site is fully consistent with the rapidly reversible nature of the binding equilibrium (see comments below) whereas rapid equilibration with a binding site on the inner mitochondrial membrane is difficult to conceive since it would require facile passage of a 100,000 mol wt protein across the outer mitochondrial membrane. Thus, it is clear that the outer mitochondrial membrane is, in fact, the principal, and perhaps exclusive, structure to which the type I and II isozymes may bind.

Felgner et al.[117] have isolated a protein from the outer membrane of rat liver mitochondria which is responsible for the binding of hexokinase; incorporation of this protein into phospholipid vesicles confers the ability to bind hexokinase. The hexokinase-binding protein was characterized as an intrinsic membrane protein having a subunit molecular weight of 31,000. Subsequently, this protein was identified as the pore-forming protein ("porin") of the outer mitochondrial membrane.[123,124] Hence, hexokinase binds specifically to the very structure through which metabolites, including ADP and ATP, traverse the outer mitochondrial membrane. This location permits mitochondrial hexokinase to play an important role in governing the interactions between a predominantly cytoplasmic glycolytic metabolism and mitochondrial oxidative phosphorylation, a concept that will be further developed later in this chapter.

D. Kinetic Properties

Many of the early kinetic studies have been reviewed by Colowick[10] and by Purich et al.[11] and, for the type I isozyme of brain, by Fromm;[125] these will not be specifically reconsidered here.

1. Mechanism

There is general agreement that the mammalian hexokinases operate by a sequential mechanism, i.e., a ternary complex of enzyme, hexose, and ATP·Mg^{++} is a required intermediate. All three isozymes from a variety of mammalian tissues[10,11,24,44,48,81,82,126,127] yield Lineweaver-Burk plots intersecting to the left of the ordinate, indicative of a sequential mechanism.[128] Whether the mechanism is, in fact, a rapid equilibrium random addition mechanism, as proposed for the type I hexokinase of brain,[11,125,129] erythrocyte hexokinase[130]

Table 3
COMPARATIVE K_m VALUES FOR ISOZYMES I TO III FROM VARIOUS MAMMALIAN TISSUES

K_m for Glucose (*M*)			K_m for ATP·Mg^{++} (*M*)			
I	II	III	I	II	III	Ref.
4×10^{-5}	1.6×10^{-4}	4×10^{-5}	4×10^{-4}	7×10^{-4}	14×10^{-4}	17
$2–3 \times 10^{-5}$	$1–2 \times 10^{-4}$	$1–5 \times 10^{-6}$	—	—	—	19
$4–5 \times 10^{-5}$	$2–3 \times 10^{-4}$	7×10^{-6}	4×10^{-4}	$7–8 \times 10^{-4}$	9.8×10^{-4}	24
6×10^{-5}	—	—	$7–12 \times 10^{-4}$	—	—	44
$4–12 \times 10^{-5}$	—	—	5×10^{-4}	—	—	47
10×10^{-5}	—	$3–4 \times 10^{-5}$	$14–16 \times 10^{-4}$	—	$14–16 \times 10^{-4}$	49
3×10^{-5}	3×10^{-4}	$5–7 \times 10^{-6}$	—	—	—	53
4×10^{-5}	1.7×10^{-4}	4×10^{-5}	4×10^{-4}	7×10^{-4}	$\sim 10 \times 10^{-4}$	57, 92
8×10^{-5}	—	—	6×10^{-4}	—	—	77
2×10^{-5}	—	—	4×10^{-4}	—	—	78
6×10^{-5}	—	—	6×10^{-4}	—	—	81
—	2×10^{-4}	—	—	11×10^{-3}	—	82
—	1.1×10^{-4}	—	—	1.7×10^{-4}	—	126
6×10^{-5}	1.4×10^{-4}	1.5×10^{-5}	5×10^{-4}	3.5×10^{-4}	4.5×10^{-4}	127
$3–5 \times 10^{-5}$	—	—	2×10^{-4}	—	—	133

(isozyme unspecified, but presumably type I), and possibly the type II isozyme of skeletal muscle also,[126] remains uncertain.[12,131,132a-b]

2. Apparent Affinity for Substrates

The kinetically determined K_m value may or may not be equivalent to a thermodynamic dissociation constant, K_d, depending on the particular kinetic mechanism applicable to the enzyme in question.[128] In the case of hexokinase, it has not been established that equating K_m values with K_d is appropriate.[12,131,132a-b] However, K_m values do provide a useful measure of apparent affinity, and it is in this manner that we will use them here. It is also well known that K_m values may be dependent upon the conditions (pH, temperature, etc.) under which they are determined and these conditions have not been uniform in all the published studies considered here. Nevertheless, the reported values for the isozymes from a variety of mammalian tissues are remarkably similar (Table 3). Although there may be a few exceptions, the apparent affinity of the isozymes for glucose increases in the order II<I<III. Several studies[24,49,57,92,127] have confirmed the early observation[17,19] that only the type III isozyme shows substrate inhibition at glucose concentrations above approximately 0.3 m*M*. All of the isozymes can phosphorylate sugars other than glucose, usually with considerably higher K_m values, e.g., K_m values for fructose are generally well in excess of 1 m*M*.[17,19,24,44,47,77,78,81,92,127] The isozymes are all quite similar in their K_m for ATP·Mg^{++}, with values of approximately 0.5 m*M* being most frequently reported.

3. Inhibition

ADP (or ADP·Mg^{++}) has been reported to be a noncompetitive inhibitor vs. both glucose and ATP·Mg^{++} with the type I isozyme from brain[11,24,125] and ascites tumor,[134] the type II isozyme from skeletal muscle[24,126] and ascites tumor,[134] and the type III isozyme from rat liver;[24] with the type I isozyme from porcine heart, however, ADP·Mg^{++} was found to be noncompetitive vs. glucose but competitive vs. ATP·Mg^{++}. Noncompetitive inhibition by ADP is inconsistent with a simple rapid equilibrium random addition mechanism, leading Fromm and colleagues[11,125] to propose a modification of this mechanism, including the existence of a second (inhibitory) allosteric site for binding ADP; the observation that the inhibition exhibited greater than first-order dependence on ADP concentration was also considered to support the postulated existence of a second site. However, there is evidence against the existence of a second site,[12,132a] and it seems quite possible that the observed ADP inhibition pattern may be explained in a different manner and with a mechanism involving a single nucleotide binding site, e.g., the mechanism proposed by Storer and Cornish-Bowden for rat liver glucokinase which, like hexokinase, shows noncompetitive inhibition by ADP or ADP·Mg^{++} vs. both substrates, with a greater than first-order dependence on ADP concentration.[32] Leaving aside the obviously controversial explanation for the observed inhibition pattern, it is clear that inhibition by ADP (or ADP·Mg^{++}), with K_i values of approximately 1 m*M*,[11,24,44,126] is a common feature of all three isozymes. At least with the brain (type I) isozyme, other nucleotides (e.g., AMP, GDP, UDP) also inhibit the enzyme with K_i values greatly in excess of 1 m*M*.[135]

Glucose-6-P is a rather potent inhibitor of all three isozymes of hexokinase, with K_i values in the low μ*M* range.[10-12,24,126] Inhibition was found to be competitive vs. ATP·Mg^{++} with the type I[11,24] and III[24] isozymes, but either competitive[24] or noncompetitive[126] with the type II isozyme, an apparent discrepancy that has not been resolved; glucose-6-P was noncompetitive vs. glucose with all isozymes.[11,24,126] Rose and co-workers[136,137] have observed an interesting difference between the type I and II isozymes in the *time* required for inhibition by added glucose-6-P, the type II isozyme exhibiting a pronounced lag before responding fully to glucose-6-P while inhibition of the type I isozyme was virtually instantaneous. When bound to mitochondria, the response of the type II isozyme was even more prolonged.

The mechanism by which glucose-6-P inhibits hexokinase remains controversial. The background for the current debate on this question has been provided previously[10-13,125] and will not be repeated here. However, in brief, the question is: is competitive (vs. ATP·Mg^{++}) inhibition by glucose-6-P the result of binding at a discrete allosteric site, as proposed many years ago by Crane and Sols,[138] or is it the result of a simple competition between the γ-phosphate of ATP and the 6-phosphate of glucose 6-phosphate for a common anion binding site on the enzyme as advocated by Fromm and colleagues.[11,125,139] This clearly remains an unsettled matter.[132a,139-141] The interested reader may consult Figure 1 and its legend for an explanation felt by this author to be most consistent with the available results.

Glucose-1,6-bisphosphate has also been shown to be a rather potent inhibitor of mammalian hexokinases.[48,130,138,145-149] All three isozymes are inhibited competitively (vs. ATP·Mg^{++}), with K_i values in the range of 10 to 100 μ*M*;[48,130,145-149] the reported insensitivity of the type I isozyme[148] was subsequently[147,149] attributed to the marked pH dependence of the inhibition by glucose 1,6-bisphosphate and prevention of this inhibition by excess Mg^{++} (i.e., the Mg^{++} chelate of glucose-1,6-bisphophate does not inhibit the enzyme).[146]

In view of the similarity in the inhibition by glucose-6-P and glucose-1,6-bisphosphate (competitive vs. ATP·Mg^{++}) as well as their common ability to cause solubilization of the mitochondrial hexokinase, it has been suggested that these compounds bind to a common high affinity regulatory site on the enzyme.[145] There is an excellent correlation between the ability of various hexose 6-phosphates to inhibit the enzyme[138] and to solubilize the mitochondrial hexokinase,[97,145,150] and the ability of these compounds to induce conformational change in the molecule.[131,144] Thus, both inhibition and solubilization appear to result from conformational changes induced by the binding of glucose-6-P or glucose-1,6-bisphosphate to a regulatory site on hexokinase (Figure 1).

Another similarity between the inhibition by glucose-6-P and glucose-1,6-bisphosphate is that, for the type I isozyme, this inhibition is antagonized by P_i.[11,46,48,60,101,125,137,139,142,143,147] In contrast, inhibition of the type II isozyme is *not* relieved by P_i.[125,126,137] This difference in the effect of P_i on inhibition by glucose-6-P (or glucose-1,6-bisphosphate) is perhaps the

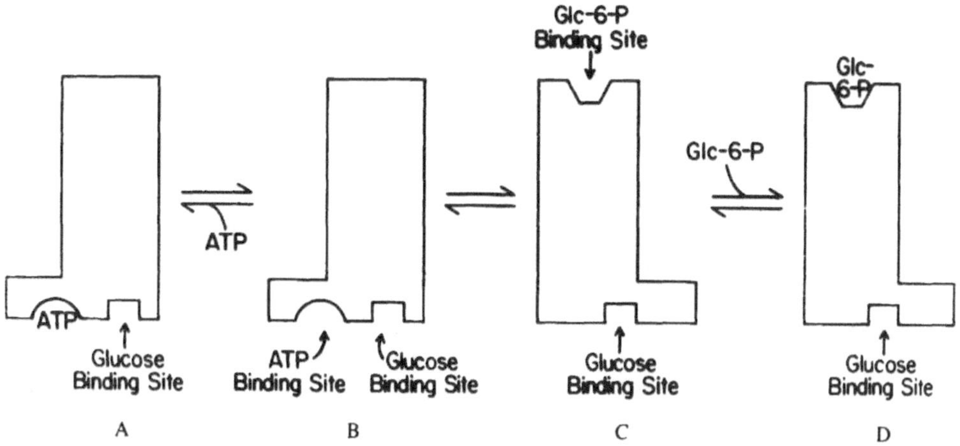

FIGURE 1. Interactions between an allosteric glucose 6-phosphate binding site and the nucleotide binding site of brain (type I) hexokinase. The enzyme is considered to exist in (at least) two conformations (central figures, B and C). One conformation (left center, B) is, in the absence of added ligands, the preferred conformation and can bind ATP at the catalytic site but is not capable of binding glucose-6-P at an allosteric site. The other conformation (right center, C) can bind glucose-6-P with *high* affinity at an allosteric site, but has lost the ability to bind ATP, and is thus inactive. Both forms possess a glucose binding site to which glucose-6-P can also bind with *low* affinity. It is assumed that glucose-6-P and ATP cannot simultaneously bind to the active form (on the left) due to overlap of the γ-phosphate of ATP and the 6-phosphate of glucose-6-P, in accordance with the thoughts of Fromm and co-workers;[11,125,139] however, prevention of ATP binding in this manner is not a major factor at relatively low (<1 m*M*) levels of glucose-6-P due to the low affinity for binding of glucose-6-P to this conformation. It is evident that the equilibrium would be shifted toward the active or inactive form by preferential binding of ATP or glucose-6-P, respectively, thereby accounting for the competitive (vs. ATP) inhibition by glucose-6-P[11,24] as well as the mutually exclusive binding of these ligands.[142] It is also evident that formation of an enzyme·glucose·glucose-6-P complex[131,143] is compatible with this model. Other observations which led to the development of this proposal include: (1) an excellent correlation between the ability of hexose 6-phosphates to inhibit brain hexokinase and their ability to induce conformational change in the molecule[131,144] strongly suggesting that the latter is required for inhibition, (2) an excellent correlation between the ability of hexose 6-phosphates to inhibit brain hexokinase and their ability to inhibit the binding of tetraiodofluorescein, a fluorescent nucleotide analog which is a competitive inhibitor (vs. ATP) of the enzyme;[132a] hence, inhibitory hexose 6-phosphates appear to influence the structure of the enzyme at the site to which tetraiodofluorescein (or an aromatic nucleotide ring) binds with resulting enhancement of fluorescence, (3) both kinetic results[141] as well as conventional binding studies[140] indicate the existence of *two* binding sites for glucose-6-P. The latter indicate that binding of glucose-6-P to the high affinity site is unaffected by *N*-acetylglucosamine, a competitive inhibitor (vs. glucose) of the enzyme,[133] while binding at the low affinity site is antagonized by *N*-acetylglucosamine.[140] 1,5-Anhydroglucitol-6-P, a glucose-6-P analog which is also a potent inhibitor of the enzyme,[138] effectively inhibits binding of glucose-6-P to the high affinity site while glucose actually enhances binding.[140,143] These observations clearly indicate that the high affinity glucose-6-P binding site must be an inhibitory site discrete from the catalytic site to which glucose or *N*-acetylglucosamine may bind.

most striking kinetic difference between the type I and II isozymes. The effect of P_i on inhibition of the type III isozyme has apparently not been examined.

The type I and II isozymes also differ in their response to P_i itself. In the absence of added glucose-6-P, physiologically reasonable levels of P_i have no effect on the type I isozyme,[10-12,125,137,142,143] although inhibition has been observed at much higher concentrations (K_i = 20 to 35 m*M*).[44,143,151] The type II isozyme from rat skeletal muscle was rather sensitive to inhibition by P_i ($K_i \simeq 3$ m*M*),[126] but the type II isozyme from ascites tumor cells was not inhibited by P_i concentrations up to 4 m*M*.[137] Studies of the type II isozyme from other sources, as well as the type III isozyme, are necessary before any generalizations can be made concerning the response of these isozymes to P_i.

As discussed above, it has been observed that the inhibition of type I hexokinase from many sources by glucose-6-P is reversed by P_i. Two studies have suggested that the type I isozyme from pig heart may be an exception.[44,152] Vowles and Easterby[44] found that P_i did not relieve the inhibition of the type I heart hexokinase by glucose-6-P, while Shone and Fromm[152] showed that the type I isozyme from pig brain, like the type I isozyme from other tissues, did respond to P_i. Thus, it was suggested[152] that pig brain and heart might contain distinct "type I" isozymes. There have also been reports that the type I isozyme, or subforms thereof, from human blood cells might respond poorly if at all, to P_i in reversal of inhibition by glucose-6-P or glucose-1,6-bisphosphate,[46,60] but this is at variance with results of other investigators.[48,137] Reported differences between the various subforms in their affinity (K_i) for glucose-1,6-bisphosphate and their relative sensitivity to phosphate reversal of inhibition by the bisphosphate[46,153a] have also not been confirmed.[46,48] Particularly in view of the possible artifactual nature of the subforms (see Section II.C), further consideration should be given to the possibility that the discrepancies in the reported responses of the erythrocyte enzymes to P_i[46,48,60,137,153a] might be due to post-lytic modifications affecting the kinetic properties of the enzyme. This is even more the case with the heart enzyme since it has already been reported[101] that the response to P_i in reversing inhibition by glucose-6-P can be markedly decreased in "aged" preparations of the partially purified hexokinase from dog (but not guinea pig) heart, clearly indicating that artifactual loss of the ability to respond to P_i can occur. Recently, a preliminary report by Easterby and Qadri[153b] indicates that this is also the case with the pig heart enzyme. Thus, the reported differences between the type I isozymes from various sources in their response to P_i in reversal of inhibition by glucose-6-P may not be representative of the *in situ* condition. The apparent ease with which this response to P_i can be lost without other notable effect on the enzyme is quite intriguing since it suggests that the P_i response is associated with a rather labile structural feature of the molecule that is not, apparently, otherwise involved in catalytic function.

While discussing inhibition of hexokinase, it is of interest to mention inhibition by a nonphysiological compound that is of clinical relevance. Sols and Crane[154] conducted an extensive study of the substrate specificity of the type I hexokinase from brain. One of their conclusions was that appreciable structural variation at carbon 2 of the glucose molecule could be tolerated without marked effect on the ability of the hexose to interact with the enzyme; thus, 2-deoxyglucose, mannose (epimeric with glucose at C-2), as well as glucosamine (amino function replaces hydroxyl at C-2) and *N*-acetylglucosamine are all either substrates or competitive (vs. glucose) inhibitors of the enzyme. Bertoni[155,156] has pointed out that metrizamide, widely used as a radiological contrast agent, is an N-substituted glucosamine derivative that, based on the work of Sols and Crane,[154] may be expected to inhibit brain hexokinase, and has suggested that such inhibition may account for at least some of the effects observed in patients receiving metrizamide. Bertoni[155,156] has shown that metrizamide does serve as a competitive (vs. glucose) inhibitor of brain hexokinase from several species, including humans; the K_i (about 2.5 m*M*) with human brain hexokinase[156] is well below estimated concentrations (about 50 m*M*) of metrizamide in physiological fluids of patients receiving this agent and hence inhibition could indeed be reasonably expected.

Studies on the substrate specificity of the type II and III isozymes, comparable to the extensive study of the type I isozyme by Sols and Crane,[154] apparently have not been done. However, as noted above, several investigators have shown that all three isozymes can phosphorylate sugars other than glucose, and Grossbard and Schimke[24] have found the type I, II, and III isozymes from rat tissues to be quite similar in their susceptibility to inhibition by various sugars, suggesting a corresponding similarity in the specificity of the three isozymes for different glucose analogs. Taken together with the many other structural and kinetic similarities of the isozymes discussed earlier, it seems likely that the type II and III isozymes also would be inhibited by metrizamide. Hence, this agent may be disruptive of carbohydrate metabolism in other tissues (possibly containing other isozymes) in addition to brain.

4. Nature of the Catalytic Site

Sols and Crane[154] had reported that there was no detectable difference in the rate with which the α- and β-anomers of glucose were phosphorylated by brain (type I) hexokinase. However, more recent studies by Okuda et al.[157] have indicated that the type I enzyme from several rat tissues exhibited V_{max} values with the β-anomer that were approximately 1.5 times larger than those with the α-anomer. A similar ratio in V_{max} values was seen with the type II and III isozymes. Since no standard deviations were indicated, it is not possible to assess whether the slight differences between the V_{max} ratios (β/α) for the different isozymes, or between the same isozyme from different tissues, were significant. In any case, based on these results, it would appear that all three isozymes are similar in preferring the β-anomer, with V_{max} ratios (β/α) of about 1.5.[157] These results, coupled with an overall similarity in substrate specificity (Section III.D.3) suggest a corresponding structural similarity in the hexose binding sites of the three isozymes.

The isozymes are also virtually identical in their specificity for ATP as a phosphoryl donor, with all three isozymes having activities only 4 to 8% that seen with ATP when ITP was substituted as phosphoryl donor; other nucleoside triphosphates were used even less effectively.[24] Darby and Trayer[158] have recently defined the stereospecificity of the metal-nucleotide binding site on type II hexokinase from rat muscle.

Both the type I[159] and type II[160,161] isozymes are inactivated by glucose-derived alkylating agents, and in both cases, derivatization of a single cysteine residue, presumably located at the catalytic site, resulted in complete inactivation of the enzyme. With the type I enzyme, a second cysteine residue, not required for activity, also reacted with the alkylating glucose analog.[159] Studies with an ATP analog, reactive with sulfhydryl groups, led Subbarao and Kenkare[162] to suggest that the catalytic site of type I hexokinase from bovine brain contained two sulfhydryl groups which were in close proximity and only one of which was essential for catalytic activity. The catalytic importance of sulfhydryls in brain hexokinase, and the marked effect of ligand binding on the reactivity of these sulfhydryls, has been extensively studied,[75,163,164] and it has even been suggested[165] that the activity of erythrocyte hexokinase might be regulated by the oxidation status of critical sulfhydryls in the enzyme as governed by the cellular redox state.

Solheim and Fromm[166] have studied the pH dependence of bovine brain hexokinase and obtained evidence for the involvement of histidyl and glutamate or aspartate residues in catalytic function. Arginine residues are required for catalysis by the type I isozyme from rat brain.[167] Aside from the critical sulfhydryl residue,[160,161] the nature of the amino acids required for catalytic function of the type II isozyme is unknown. There appears to be no information at all on critical residues in the type III isozyme. The many other similarities of these isozymes suggests, however, that similar mechanisms involving these same functionally important residues may be found for all three. Observations suggesting that the major differences between the isozymes seem to be located in surface regions of the molecule (Section III.B), rather than more internal regions likely to be most directly involved in catalysis,[168] would be consistent with the view that the basic structural features of the catalytic site are highly conserved in the mammalian hexokinases.

IV. CONSEQUENCES OF THE BINDING OF HEXOKINASE TO MITOCHONDRIA

Both type I and type II hexokinase can bind to the outer mitochondrial membrane, and more specifically, to the structural pore through which metabolites gain access to and egress from the mitochondria (Section III.C). Oxidative phosphorylation, a metabolic characteristic of the mitochondria, is by far the major source of the ATP produced in aerobic mammalian cells; this process, of course, results in the conversion of ADP to ATP. On the other hand, the major ATP utilizing reactions (e.g., biosynthetic activities, ion movements at the plasma membrane) are predominantly extramitochondrial; these reactions result in conversion of ATP to ADP (or, in a few cases, AMP, which due to the widespread occurrence of adenylate kinase activity, can be considered equivalent to ADP). Therefore, it is apparent that normal functioning of aerobic mammalian cells requires an effective balance between the supply of ADP for oxidative phosphorylation and the provision of ATP for cellular energy requirements. Oxidative phosphorylation also requires a supply of oxidizable substrate which, in many cells, is pyruvate derived from glycolytic metabolism of glucose. And, of course, the initial step in utilization of glucose is its phosphorylation by hexokinase, resulting in concomitant conversion of ATP to ADP. Physical location of hexokinase at the mitochondrial pore clearly suggests the possibility that this enzyme may effectively function in governing the flux of ADP and ATP between mitochondrial and extramitochondrial compartments, and thereby result in integrated control of the initial step of glucose metabolism and the terminal stages, represented by oxidative phosphorylation. The interested reader may consult Bessman and Geiger,[14] Rose and Warms,[97] and Moore and Jöbsis,[169] as well as the somewhat prophetic article by Siekevitz [170] for additional relevant comments.

A. Preferential Utilization of Mitochondrially Generated ATP by Mitochondrially Bound Hexokinase

Rose and Warms[97] examined the possibility that a mitochondrial location of hexokinase might facilitate access to ATP generated intramitochondrially by oxidative phosphorylation, but were unable to find evidence for this in experiments involving competition between mitochondrial hexokinase and added yeast glycerol kinase for ATP generated intramitochondrially by oxidative phosphorylation. In contrast, Felgner[171] reported that similar competition experiments indicated that the mitochondrially bound hexokinase did have preferential access to intramitochondrially generated ATP but a detailed account of this work has not been published. Recently, Schlame et al.[122] have reported that mitochondrially bound hexokinase of reticulocytes does not preferentially utilize intramitochondrially generated ATP. However, these workers used mitochondria that had been extensively depleted (approximately 13% of the original activity remained) of hexokinase by repeated incubation with glucose-6-P; the remaining "tightly bound" hexokinase was suggested to be bound to the inner mitochondrial membrane although, in fact, the location of this enzyme, and even its association with the mitochondria, was not demonstrated. Thus, the significance of these results[122] for interpreting the physiological function of what is by far the major portion of the mitochondrial hexokinase, bound to the outer membrane, is problematic.

In contrast to the somewhat equivocal results discussed above, several investigators have produced rather clearcut evidence that the mitochondrial hexokinase of liver,[14,172,173] diaphragm,[14,173] heart,[14,174] and skeletal muscle[173] preferentially utilizes intramitochondrially generated ATP. In addition, Kurokawa et al.[175] have shown that mitochondrially bound hexokinase is much more effective at utilizing intramitochondrially generated ATP than is the nonbound hexokinase.

ADP produced by the mitochondrial hexokinase reaction might also be expected to have ready access to the oxidative phosphorylation system. That this is the case has been shown by several workers who demonstrated marked stimulation of oxygen uptake by addition of glucose to hexokinase-bearing mitochondria from brain,[169] ascites tumor cells,[176] or rat hepatoma.[177] In each case the direct involvement of the ADP-generating hexokinase reaction was convincingly established.

Thus, the mitochondrial location of hexokinase results in definition of a discrete cellular microenvironment in which the ADP/ATP levels are directly related to the functionally opposite (i.e., ATP producing, ADP consuming vs. ATP consuming, ADP producing) metabolic reactions of oxidative phosphorylation and glucose phosphorylation. The net result

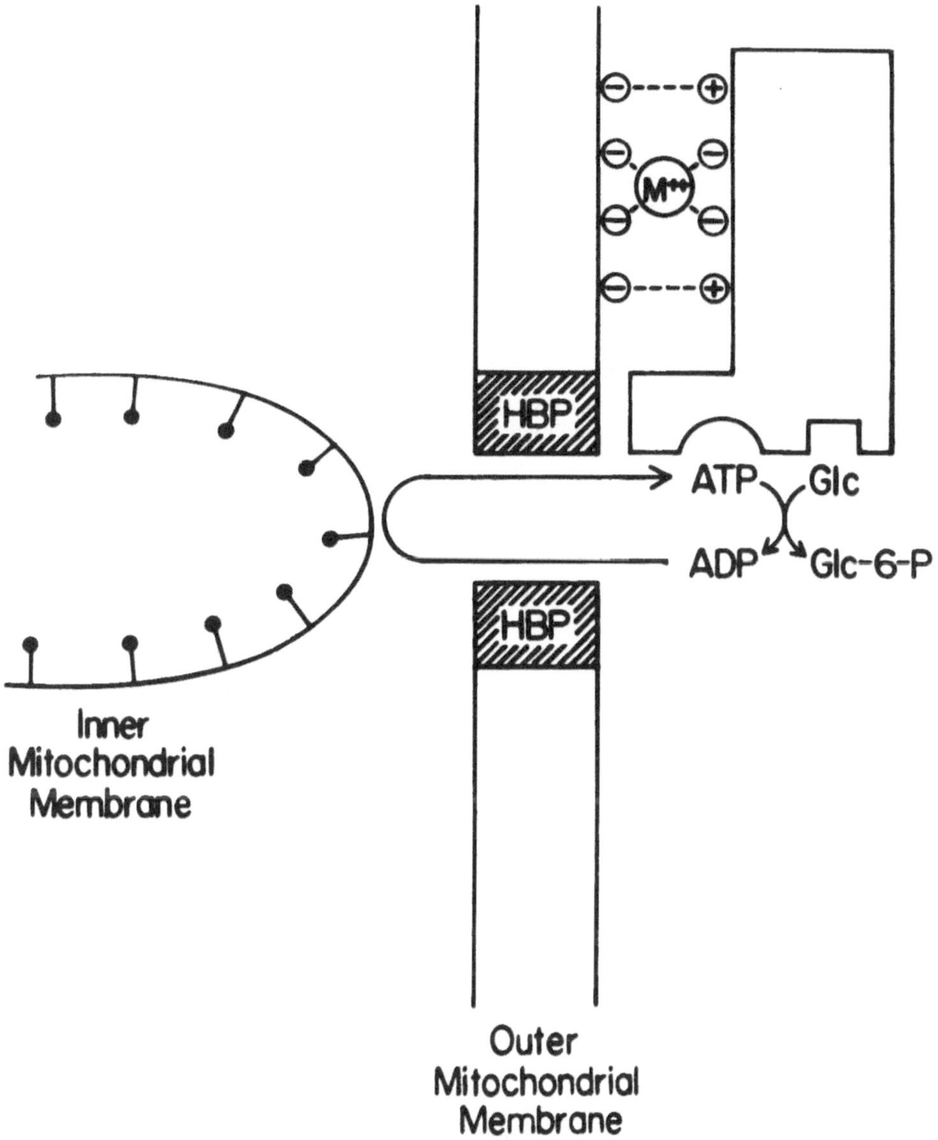

FIGURE 2. Topological relationship between mitochondrial hexokinase and oxidative phosphorylation. In the absence of glucose-6-P binding to the allosteric site (Figure 1), hexokinase exists in a preferred conformation capable of bimodal interactions with the outer mitochondrial membrane. Divalent cations provide a bridge between negatively charged groups on enzyme and membrane.[112] In addition, the enzyme binds specifically[117] to the pore structure[123,124] through which metabolites traverse the outer mitochondrial membrane. The enzyme is therefore in a position to effectively utilize ATP generated intramitochondrially by oxidative phosphorylation, while the ADP produced by the hexokinase reaction has ready access to the oxidative phosphorylation system of the inner mitochondrial membrane. This spatial relationship provides a means for effectively coupling the initial step of glucose utilization with the terminal stage of that process, represented by the oxidative reactions occurring in the mitochondria. HBP, hexokinase binding protein.

of this is that stimulation of *either* reaction can result in a coordinated increase in activity of the other, i.e., an integrated control of the initial and terminal stages of what is a major energy generating pathway in many mammalian cells. The special relationship between oxidative phosphorylation and glucose phosphorylation catalyzed by mitochondrially bound hexokinase is represented in Figure 2.

B. Effect of Binding on the Kinetic Properties of Hexokinase

Several studies have indicated that the apparent affinity for ATP is increased (i.e., K_m is decreased) as a result of binding of hexokinase to the mitochondria.[56,112,113,175,178] Though the differences were not large, generally on the order of two- to fivefold, this could impart a significant kinetic advantage to the bound enzyme in competing for available ATP, as has been suggested.[12,13,97,113] Perhaps more important in this respect, however, is the preferential access to intramitochondrially generated ATP (discussed above) which is enjoyed by the mitochondrial enzyme.

It has also been reported that the susceptibility of hexokinase to inhibition by glucose-6-P is decreased by binding to the mitochondria, i.e., the K_i for glucose-6-P is increased by binding.[56,113,178] In view of the close similarity between inhibition by glucose-6-P and by glucose-1,6-bisphosphate, it may be expected that similar differences would be seen between the K_i values of the soluble and bound enzymes for glucose-1,6-bisphosphate, though this has apparently not been determined directly. Increased apparent affinity for ATP, coupled with decreased susceptibility to inhibition by glucose-6-P (and, presumably, glucose-1,6-bisphosphate) could lead to substantial differences in activity, the bound form being the more active.

C. Mitochondrial Hexokinase and Glucose Utilization

Based on the logic outlined above, binding of hexokinase to the mitochondria confers specific kinetic advantages that should make the enzyme highly effective at initiating glucose metabolism as well as being conducive to the coordinated regulation of glucose metabolism and oxidative phosphorylation, processes which are in turn dependent on cellular energy status.[179] As such, we could expect mitochondrial hexokinase to figure prominently in the metabolism of cells which exhibit a major dependence on blood-borne glucose as a substrate and in which the rate of glucose metabolism is rapidly responsive to altered requirements for metabolic energy. There is considerable evidence to support the view that this is not simply a theoretically advantageous system, but rather one that is indeed exploited in the functioning living cell.

Before discussing some examples of this, it is appropriate to restate a caveat mentioned earlier (Section III.A), i.e., there is no reason to believe that the "averaged" metabolic properties of an entire tissue (or tissue homogenate) are necessarily characteristic of all cells within what may be a quite heterogeneous collection of cell types comprising the tissue. Since, in most cases, our information is based on studies with the entire tissue (e.g., "liver", "brain", etc.), any inferences made from these studies should be considered with this caveat in mind. In other words, attempting to relate specific metabolic properties to specific enzymatic features based on studies with heterogeneous systems is necessarily tenuous. However, it is frequently the only information we have and can be instructive. Fortunately, there is also information available from more homogeneous systems to supplement that from "whole tissue" studies.

Felgner and Wilson[70] considered blood-borne glucose and glycogen as alternative sources for glucose-6-P entering glycolytic metabolism. They pointed out that tissue glycogen contents were generally correlated with the levels of enzymes involved in glycogen synthesis or degradation,[180] and chose phosphoglucomutase as an enzyme representative of the capacity of the tissue for glycogen metabolism. Mitochondria isolated from various rat tissues were then examined for their relative hexokinase content; this differed greatly in the mitochondria derived from different tissues although other properties such as the relative cytochrome content were very similar. An inverse relation was observed between tissue phosphoglucomutase levels and the relative hexokinase content of the mitochondria. Felgner and Wilson[70] suggested that, in tissues exhibiting less dependence on glycogen (i.e., low phosphoglucomutase levels) and thus greater dependence on blood glucose as a source of glucose-6-P,

the mitochondria were "tailored" for effective glucose phosphorylation by having increased mitochondrial hexokinase activity.

Ballatori and Cohen[66] assayed both total hexokinase activity and mitochondrial hexokinase activity in dissected cortical, medullary, and papillary regions of rat kidney. They observed that total hexokinase activity (units per gram tissue basis) increased in the order cortex $<$ medulla $<$ papilla, there being an approximately twofold difference between cortex and papilla. This could be correlated with the relative rate of glycolytic glucose metabolism, which was much higher in papilla than in cortex or medulla. More striking, however, was the finding by Ballatori and Cohen[66] that the specific activity of hexokinase in the mitochondrial fraction from the papilla was approximately sixfold higher than the specific activity of hexokinase in cortical mitochondria. Thus, in dissected kidney regions, there again appears to be "tailoring" of mitochondria such that increased mitochondrial hexokinase content is associated with increased rates of metabolism of blood-borne glucose.

Synaptosomes are vesicular structures that are formed by "pinching off" nerve terminals during homogenization of neural tissue.[181] Nerve terminals are thought to be responsible for utilization of the major portion of the total glucose requirement of brain,[182-184] and it has been shown that nerve terminal mitochondria are particularly high in hexokinase content.[99] Actually, even the mitochondria derived from non-nerve terminal fractions of brain[99] are very high in hexokinase content relative to mitochondria from other tissues,[70] reflecting the well-known virtually total dependence of the brain (under normal conditions) on blood-borne glucose as a substrate for what is an extremely high rate of energy metabolism,[12,13] rapidly responsive to functional activity.[185,186]

The rate of glucose utilization by mammary gland is greatly increased during lactation, providing metabolic energy, reducing equivalents (i.e., NADPH via the hexose monophosphate pathway), as well as precursors for the extensive lipid synthesis that accompanies lactation.[187] Transient increases in the activities (U/g tissue) of several relevant enzymes, including a several fold increase in hexokinase, occur during lactation, with decreases to control levels accompanying post-weaning involution.[72,187,188] Kopelovich et al.[188] found that the increase in hexokinase occurred selectively in the mitochondrial fraction, the hexokinase activity of the soluble fraction actually declining slightly during lactation; the specific activity of hexokinase in mammary gland mitochondria prepared from lactating animals was more than 7-fold higher than that found in mitochondria from pregnant animals and 100-fold that of virgin mice. Hence, mitochondrial hexokinase content closely parallels the demand for glucose metabolism in support of lactation.

Maternally supplied glucose represents the major substrate for energy metabolism as well as the many biosynthetic processes of the fetus.[189] Fetal liver exhibits active glucose catabolism[190] which contrasts with the adult liver which does not[190] and which primarily utilizes noncarbohydrate substrates.[68] It is therefore of interest to note that development is accompanied not only by a decline in total hexokinase activity (U/g tissue) of the liver[36,191,192] but, more specifically, by a marked decline in the mitochondrial activity.[191,192]

The results discussed above lend support to the view that a mitochondrial location for hexokinase does indeed represent a metabolic advantage, an advantage expressed in a variety of normal mammalian tissues exhibiting a heavy reliance on glucose as a substrate for cellular metabolism. Malignant tissues are *abnormal* tissues characterized by a high rate of glucose utilization. In contrast to normal mammalian tissues, tumors metabolize a high proportion of the glucose to lactate even under aerobic conditions. This high rate of aerobic glycolytic metabolism to lactate was emphasized as a characteristic of tumors by Otto Warburg[193] and is commonly referred to as the Warburg effect.[194] The explanation for this effect and its significance for tumor metabolism remain somewhat controversial and are not directly germane to the present subject; the interested reader may consult an article by Racker and Spector[195] and the excellent comprehensive review by Pedersen.[196] It is now clear that this

active aerobic glycolytic metabolism characteristic of tumors is directly correlated[177,197,198] with the presence of high levels of mitochondrial hexokinase in these cells, as has been suggested by several investigators.[199-205] As with the mitochondrially bound enzyme of normal tissues, tumor mitochondrial hexokinase is relatively resistant to inhibition by glucose-6-P,[177] thereby permitting high rates of glucose phosphorylation (and thus glycolysis) even in the presence of relatively high cellular concentrations of this product. It is not implied here that the high levels of tumor mitochondrial hexokinase are the sole cause of the Warburg effect. Indeed, Pedersen[196] has pointed out other differences between normal tissues and tumors, one of the most relevant in the present context being markedly decreased mitochondrial content in tumors exhibiting high rates of aerobic lactate formation. Hence, increased introduction of glucose into glycolysis via the mitochondrial hexokinase reaction coupled with a decreased capacity for mitochondrial oxidation of pyruvate may lead to the high rate of lactate formation.

Several of the tumors examined in these studies were hepatomas,[197-200] and it may be useful to specifically note an important difference between the glucose metabolism in these cells and that in normal hepatocytes from which the hepatomas are presumably derived. It is clear that phosphorylation represents the rate-limiting step in glucose utilization by normal liver.[198,199,206] Since hepatocytes contain little or no hexokinase activity,[27-29] glucose phosphorylation in these cells will depend on glucokinase, and thus, due to the high K_m of this enzyme for glucose,[15-17] on serum levels of glucose. As pointed out by Weinhouse et al.,[207] a major metabolic difference between the normal hepatocyte and the malignant hepatoma is this conversion from a "low affinity" (glucokinase) to a "high affinity" (hexokinase) capability for glucose metabolism. It is conceivable that this difference, coupled with the importance of glucose metabolism for tumor growth, might be useful were it possible to target metabolic inhibitors for a specific organ.

V. REGULATION OF HEXOKINASE ACTIVITY

We have thus far discussed many of the properties of mammalian hexokinases and, in a preliminary manner, related them to regulation of in vivo activity. It is our purpose here to present a more integrated discussion of these regulatory influences, and to introduce other factors of possible significance.

A. The Influence of Metabolites on Catalytic Activity

We are concerned here with small molecules of physiological occurrence which have been shown to affect hexokinase activity. Metabolites likely to exert significant influence on enzymatic activity in vivo include nucleotides, glucose-6-P and glucose-1,6-bisphosphate, and P_i.

1. Nucleotides

As discussed in Section III.D.3, ADP has been shown to inhibit all three isozymes of mammalian hexokinase. With the type I isozyme, and possibly the others (this apparently has not been examined), other nucleotides also inhibit the enzyme. The K_i values for ADP ($\sim$ 0.6 to 6 mM)[24,135] are somewhat higher than the estimated intracellular levels of ADP in brain,[208,209] muscle,[210] and other mammalian tissues,[211] making the *assumption* that the nucleotide is uniformly distributed in tissue water. However, there is evidence that much of the cellular ADP is actually intramitochondrial,[212] and thus the concentration (if one may use that term) in the cytoplasm that is accessible to hexokinase may be much lower than the K_i values measured in vitro. Other nucleotides are present in even lower amounts than the adenine nucleotides.[211] Assuming that the rather high K_i values observed for inhibition of the type I isozyme by various nucleotides[135] are also representative of the other isozymes,

we believe it is unlikely that inhibition by nucleotides represents a significant regulatory influence on mammalian hexokinase activity in vivo.

An exception to this may be inhibition by the nonchelated form of ATP, which has been shown to be a reasonably ($K_i \simeq 1.7$ m*M*) effective inhibitor (competitive vs. $ATP{\cdot}Mg^{++}$) of the type I isozyme from brain.[135] Since ATP represents the major nucleotide present in mammalian tissues,[208-211] and since it can be calculated[213] that a significant amount of uncomplexed ATP would exist at the estimated[214] intracellular levels (about 1 m*M*) of Mg^{++}, it is possible that this may represent a negative, though probably rather minor, influence on hexokinase activity.

2. *Glucose-6-P, Glucose-1,6-Bisphosphate, and* P_i

The early observation by Weil-Malherbe and Bone[215] that brain hexokinase was rather strongly inhibited by the product, glucose-6-P, was considerably extended by Crane and Sols,[95] who observed that sensitivity to glucose-6-P was a characteristic of hexokinases from a variety of mammalian tissues and suggested the importance of this in regulation of hexokinase activity. Kipnis and Cori[216] found that increased intracellular levels of glucose-6-P resulted in decreased hexokinase activity in diaphragm. Glycolytic metabolism in the erythrocyte was studied by Rose and O'Connell,[217] who demonstrated that the rate of glucose utilization was inversely proportional to the cellular glucose-6-P content over a 40-fold range of the latter, thereby implicating glucose-6-P in regulation of red cell glycolysis. Other investigations indicated a prominent role for glucose-6-P in regulation of hexokinase activity in systems as diverse as heart,[218] ascites tumor,[219] and reconstructed (from purified enzymes) glycolytic systems.[220] These studies and others, taken in conjunction with the sensitivity ($K_i \approx 10$ μ*M*) of all three isozymes to glucose-6-P inhibition,[24] provide ample basis for the now generally accepted view that glucose-6-P represents a major influence on hexokinase activity in all mammalian tissues.

Inhibition by glucose-6-P cannot, however, be divorced from the effects of P_i which, as indicated in Section III.D.3, effectively antagonizes the negative influence of glucose-6-P on the type I but not the type II isozyme (with the effect on type III still undefined). The importance of this interplay between glucose-6-P and P_i is evident from several studies involving tissues rich in the type I isozyme.[218-222] Since the type I isozyme is of widespread occurrence and is, in fact, the predominant isozyme in many mammalian tissues (Table 1), it is appropriate to consider the mutually antagonistic effects of glucose-6-P and P_i as a major regulatory influence of general significance for mammalian hexokinase activity.

Although P_i serves to counteract the inhibition by glucose-6-P and in this sense may be considered to be an activator of the type I isozyme, it should be emphasized that this "activation" is manifest only in the presence of glucose-6-P, and can be attributed to the effectiveness of P_i in preventing the binding of glucose-6-P[143,223] and therefore the glucose-6-P induced conformational change[131,144,224] which results in inhibition (Figure 1). The only *direct* effect of P_i on type I hexokinase activity is an inhibition which occurs at unphysiologically high ($K_i \approx 20$ to 35 m*M*)[44,143,151] P_i concentrations and thus seems unlikely to be a significant factor in vivo.

It was suggested above (Section III.A) that the type I hexokinase served to provide glucose-6-P for a predominantly glycolytic energy metabolism in tissues or cells reliant on blood-borne glucose as a major substrate. We now see that the regulatory properties of the type I isozyme are well suited to such a role. As an illustration, we will discuss energy metabolism in brain, a tissue containing almost exclusively the type I isozyme (Table 1), but it should be evident that analogous comments could be made for any tissue which, like brain, depends on blood glucose as a substrate. It is well established that, at least under most conditions, there is a very tight coupling between functional activity and glucose consumption by the brain or regions thereof.[185,186] Let us start with the conditions as they exist in a basal steady-state situation, i.e., there are no extraordinary energy demands beyond those required for normal cellular maintenance. Thus energy production (glucose utilization) is in balance with energy consumption (i.e., utilization of high energy phosphate). Reflecting this balance, steady-state levels of various metabolites exist, including glucose-6-P and P_i. Many studies have shown that conditions that result in increased function and thus increased glucose utilization (e.g., localized stimulation or, in the extreme, general seizure activity[185,186]) also result in increased P_i and decreased glucose-6-P levels in brain,[208,209] while conditions resulting in decreased glucose utilization (e.g., anesthesia[185,186]) are accompanied by decreased P_i and increased glucose-6-P.[208] The reader may consult Siesjö[225] for a more complete discussion of the relationship between function and energy metabolism in the brain. Siesjö[225] as well as others[3,13,209] may also be consulted for a discussion of the metabolic relationships bringing about these reciprocal changes in glucose-6-P and P_i. For the present, it is sufficient to note that this is what occurs, and it should thus be evident that, based on the mutually antagonistic action of glucose-6-P and P_i, these changes may be expected to result in increased hexokinase activity during conditions requiring enhanced glucose utilization and, conversely, increased inhibition of hexokinase, and thus glucose consumption, during periods of decreased energy demand. Thus the regulatory features of the type I isozyme permit a rapid, appropriate, and coordinated response to changes in energy status of tissues, like brain, that exhibit a reliance on glucose-based energy metabolism.

In contrast to the type I isozyme, a notable feature of the type II hexokinase is its *insensitivity* to reversal of glucose-6-P inhibition by P_i. As pointed out by Katzen,[22] type II hexokinase predominates in insulin-sensitive tissues such as skeletal muscle (Table 1), and the tissue content of this isozyme is highly influenced by nutritional or hormonal status.[19,22,52,55,63,72,73] This clearly could not be tolerated by a tissue (e.g., the brain) which exhibited a continuing primary dependence on glucose as a substrate for energy metabolism, and thus survival, and it was suggested above (Section III.A) that the type II isozyme served to provide glucose-6-P that *may* be used for energy metabolism or alternatively, to support synthesis of energy storage forms (glycogen, lipids). Additional considerations, discussed below, suggest that the latter may be the major function of type II hexokinase. Thus, another way to look at this would be to say that the type I enzyme primarily serves a catabolic role while the type II hexokinase predominant in insulin-sensitive tissues is anabolic, or at least amphibolic, in function. Indeed, referring to insulin as "essentially an anabolic hormone," Elbrink and Bihler[1] have previously called attention to the fact that increase in insulin-stimulated glucose transport is normally associated with insulin-stimulated syntheses of energy reserves.

Of the several tissues considered in Table 1, only skeletal muscle exhibits a predominance of type II hexokinase in all species examined, with the possible exception of the human, and this tissue will be used as representative for discussing the role of type II hexokinase. Fatty acids are the primary substrate for basal energy metabolism in muscle.[67,68] It has been well established that stimulation of muscle is accompanied by a rapid activation of phosphorylase with resulting marked increase in glycogen utilization.[226-228] Thus, during periods of increased energy requirements, a carbohydrate-based metabolism supplements fatty acid oxidation, but the primary substrate is glycogen, not blood glucose whose transport into the cell may be rate limiting.[1] It has been estimated that 10% or less of the glucose-6-P entering glycolysis during stimulation of cat skeletal muscle was obtained via the hexokinase reaction.[210] On the other hand, it is clearly evident that hexokinase has a vital role to play when, after cessation of stimulation, glycogen stores must be replenished by drawing upon blood glucose as substrate. It is well known that the reactions of glycogen synthesis and glycogenolysis are highly regulated in a mutually opposing manner.[228] Thus, considering type II hexokinase to serve as part of the anabolic pathway from glucose to glycogen, we would anticipate that it should function well under conditions that prevail in resting muscle but be less active during the catabolic phase of glycogen metabolism associated with contraction. Piros and Staneloni[229] and others[210,230] have shown that the marked decrease in glycogen content that results from stimulation of muscle[226-229] is accompanied by a virtual doubling of the P_i content and a 10- to 20-fold increase in glucose-6-P, with estimated intracellular concentrations of glucose-6-P during stimulation ($\simeq$ 3 to 4 m*M*)[210,229,230] being greatly in excess of K_i values determined in vitro.[24,126] Thus, it may reasonably be expected that marked inhibition of hexokinase would occur during stimulation, and the inability of P_i to reverse inhibition of the type II isozyme would preserve the inhibition despite the rise in P_i levels. Indeed, based on the results of Lueck and Fromm,[126] the P_i would itself contribute further to the inhibition. After cessation of the stimulation, restoration of glycogen levels occurs rapidly,[229] and is accompanied by return of glucose-6-P and P_i to resting values. It is worth noting that the results of Piros and Staneloni[229] indicated that the marked decrease in glucose-6-P during recovery occurred somewhat more rapidly than did the increase in glycogen (Figure 4 of Piros and Staneloni[229] shows that after 2.5 min of recovery, glucose-6-P has decreased by ~50% from its level in the stimulated muscle while glycogen content has increased only ~10%); this would be consistent with the view that endogenous glucose-6-P levels had to be markedly reduced before the phosphorylation of glucose, supporting the synthesis of glycogen, could proceed at a substantial rate.

While the type I and type II isozymes catalyze the identical reaction, their properties are such that they are adapted for quite different metabolic roles, the type I being primarily catabolic and type II primarily anabolic in function. The relatively limited information available about the properties of the type III isozyme preclude any similar generalization concerning the metabolic role of this isozyme. This is further hindered by the more restricted tissue distribution (Table 1) of this isozyme, thereby making it difficult to develop any correlation between type III hexokinase content and tissue characteristics. It is clear, however, that like the other isozymes, type III hexokinase is highly sensitive to inhibition by glucose-6-P[24] and it is a reasonable expectation that this will be a major factor in regulation of its activity; the possible influence of P_i in modulating this inhibition remains to be determined. To this author, it seems unlikely that the marked substrate (glucose) inhibition,[24] uniquely exhibited by this isozyme, is a purely fortuitous property having no physiological significance. However, at the present time, the metabolic advantages that might be offered by such a response do not seem evident.

Glucose-1,6-bisphosphate is, like glucose-6-P, a relatively potent inhibitor of all three isozymes of mammalian hexokinase, and the reported K_i values[38,130,145-149] are well within the range of concentrations estimated to exist in various mammalian tissues.[231] There is therefore good reason to believe that glucose-1,6-bisphosphate may, like glucose-6-P, play an important role in regulation of hexokinase activity. As discussed above (Section III.D.3), there is considerable similarity between glucose-6-P and the bisphosphate, including their inhibition (reversible by P_i with type I isozyme) and their ability to cause release of the mitochondrial enzyme from the outer mitochondrial membrane. There are several considerations which suggest that the similarity ends there and that glucose-6-P and glucose-1,6-bisphosphate should best be regarded as metabolically discrete regulatory agents.

Ischemia results in more-or-less parallel decreases in brain levels of both glucose-6-P and glucose-1,6-bisphosphate,[231] and, coupled with the concurrent rise in P_i,[208] these changes would contribute to the increased activity of the type I isozyme that occurs during ischemia.[208] However, it is not always the case that these compounds vary in such a similar manner. Thus, in mice injected with Li^+, a 70 to 80% decrease in brain glucose-1,6-bisphosphate content was seen while glucose-6-P decreased considerably less, and it was suggested that a resulting decrease in the inhibition of hexokinase was responsible for a Li^+-induced increase in glucose utilization.[232] An even more striking difference was observed by Passonneau et al.[231] who, in agreement with results discussed earlier, found that glucose-6-P levels were greatly increased by electrical stimulation of skeletal muscle; glucose-1,6-bisphosphate levels, however, were unchanged by stimulation. In contrast, Beitner et al.[233] reported that epinephrine caused rapid increases in glucose-1,6-bisphosphate content of rat diaphragm while having no effect on glucose-6-P. Again using rat diaphragm, Beitner and colleagues[234] found that anoxia caused a decrease in glucose-1,6-bisphosphate levels while glucose-6-P, during this same time, was unchanged or even increased. Thus, it is clear that the tissue contents of glucose-6-P and glucose-1,6-bisphosphate, in many cases, vary quite independently and may therefore serve as regulators of hexokinase activity which respond differently to altered metabolic status.

Although glucose-6-P and glucose 1,6-bisphosphate are very similar in their inhibition of hexokinase and the manner in which this is achieved (see discussion above), a major difference between these compounds is in their effect on other enzymes and this is perhaps the key to understanding the physiological roles played by these compounds. In contrast to the selective action of glucose-6-P on hexokinase, the bisphosphate is known to serve as activator or inhibitor of several different enzymes relevant to carbohydrate metabolism.[3,148,149] Thus, while changes in glucose-6-P levels (or, in the case of the type I isozyme, P_i levels) may result in a rather specific influence on the hexokinase reaction and hence glucose-6-P formation, changes in glucose-1,6-bisphosphate may have a much more pervasive effect on carbohydrate metabolism, including both the formation and metabolic fate of glucose-6-P. It is not germane to the present article to discuss these more general effects of glucose 1,6-bisphosphate, but they have received explicit consideration from Beitner.[149]

B. Adenylate Energy Charge

Noting the prominent role for the adenine nucleotides in cellular energy metabolism, Atkinson[235] suggested the adenylate energy charge (EC), defined by

$$EC = \frac{1}{2}\,\frac{2[ATP] + [ADP]}{[ATP] + [ADP] + [AMP]}$$

as a parameter having wide significance for regulation of metabolic activity. Briefly, it was proposed that energy (ATP) utilizing pathways would contain enzymes responsive to the EC, with increasing activity at higher EC, i.e., when there was relatively abundant energy available in the form of high energy phosphoryl groups in the adenine nucleotide pool. Conversely, metabolic sequences generating ATP would contain enzymes responding with increased activity at lower EC, resulting in increased flux through pathways serving to replenish the high energy phosphates of the adenylates. The net result of these opposing responses was to be a stabilization of the EC at a value of $\sim$0.8 to 0.9, presumably characteristic of the stable metabolic state. Despite criticism[212,236-238] directed at both theoretical and practical aspects of the adenylate EC proposal, it has been defended[239] and has continued to receive attention from many investigators.[240]

It is not the author's intention to enter into the debate on the usefulness or validity of the EC concept. However, recognizing the interest of many investigators in EC as a possible regulatory parameter, it seems appropriate to include in this chapter some comments on the response of mammalian hexokinases to EC. Purich and Fromm[236,237] examined the relationship between EC and activity of the (type I) hexokinase from brain and erythrocytes under several differing conditions including changes in Mg^{++} concentration and addition of inhibitors such as glucose-6-P, GTP, and UTP. Although the actual responses (i.e., activity vs. EC) were, as might be expected, considerably different under the various conditions, they were generally similar in one respect — under almost all conditions examined, hexokinase activity *increased* at higher values of energy charge. It will be recognized that this is just the opposite of what would be predicted[235] for an enzyme functioning in an energy generating pathway, and it is relevant to note that the expected responses (i.e., *decreased* activity with increasing energy charge) were observed with two other enzymes of importance to glycolytic metabolism, pyruvate kinase[236,237] and phosphofructokinase.[237] Southard and Hultin[241] assessed the response of chicken skeletal muscle hexokinase (predominantly the type II isozyme) and, like Purich and Fromm,[236,237] found increased activity at higher values of EC; binding of the enzyme to mitochondria caused a slight shift of the activity vs. EC curve toward higher EC, but did not affect the overall response. The experiments of Southard and Hultin were performed under conditions of excess Mg^{++} (20 m*M* $MgCl_2$ with total adenylates of 6 m*M*) and thus can most validly be compared with results in Figure 3 of Purich and Fromm,[236] which were also obtained with excess Mg^{++}. This comparison reveals a rather remarkable similarity in response of the type I[236] and type II[241] isozymes.

It is evident that the significance of EC as a fundamental parameter governing metabolism remains controversial. Nevertheless, it is of interest that both the type I and type II isozymes behave in a manner that is, in terms of the EC concept, generally characteristic of *biosynthetic* enzymes, although the response to increasing EC was considerably more gradual and the activity at low EC considerably greater than that normally associated with biosynthetic enzymes.[235] If hexokinase is thought of as an enzyme of the energy producing glycolytic pathway, these results may seem anomalous. On the other hand, they might also be taken as a reflection of the multiple roles, both catabolic and anabolic, that the product of the hexokinase reaction, glucose-6-P, can play in cellular metabolism (discussed in the introduction to this chapter). Still another interpretation, of course, might be that the EC is not of significance as a metabolic regulatory parameter and hence the response of hexokinase is irrelevant and need not be explicable in terms of metabolic function.

Whatever view is favored by the reader, it should be emphasized that the response of hexokinase to physiologically reasonable levels of glucose-6-P, glucose-1,6-bisphosphate, and P_i is so marked that these metabolites must surely be accorded a prominent, and perhaps the most prominent, role in regulation of hexokinase activity in vivo.

C. Reversible Interactions with the Mitochondrial Membrane

It has been pointed out earlier in this chapter (Section IV) that binding of hexokinase results in the enzyme gaining privileged access to intramitochondrially generated ATP and also results in certain kinetic advantages, particularly decreased susceptibility to inhibition by the product, glucose-6-P (and probably glucose-1,6-diphosphate). It has also been noted that glucose-6-P (and glucose-1,6-bisphosphate) causes release of the enzyme from the mitochondrial membrane, an effect that is, for the type I isozyme, antagonized by P_i. Such observations led to the proposal[113] that the type I isozyme of brain existed in vivo in an equilibrium between mitochondrially bound and cytoplasmic locations, that the bound enzyme represented a more active form of the enzyme, and that the distribution of the hexokinase between the soluble and bound forms and hence hexokinase activity was responsive to cellular levels of metabolites capable of influencing that equilibrium, chiefly glucose-6-P and P_i. In other words, variation in the intracellular distribution of hexokinase is a factor in regulation of catalytic activity — hexokinase is an "ambiquitous" enzyme.[242] This proposal has been extensively discussed,[12] a discussion that will not be repeated here. There is now considerable evidence to support the proposal that hexokinase distribution *is* influenced by cellular metabolite levels which are in turn responsive to altered energy status, and that the changes are in the direction predicted, i.e., increased mitochondrially bound hexokinase, representing an activated form of the enzyme, is found during periods of increased glycolytic metabolism. Thus, ischemia, which is known to cause increased glucose utilization,[208] results in a rapid redistribution of hexokinase, with increased proportions of the total activity being found in the particulate fraction, and corresponding decreases in the soluble fraction, of

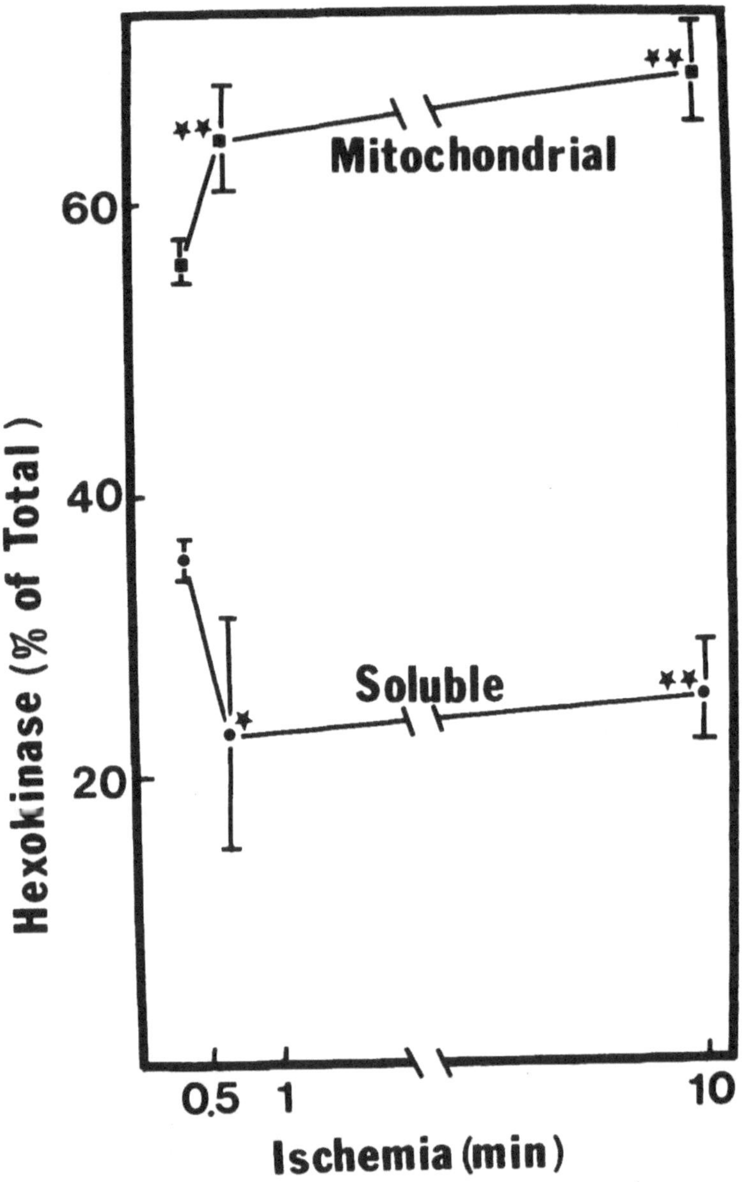

FIGURE 3. Changes in the intracellular distribution of hexokinase in the ischemic rat brain. At the indicated times after initiation of ischemia by decapitation, the brains were quickly removed and frozen in liquid nitrogen. The frozen brains were homogenized directly in ice cold 0.25 *M* sucrose, and the soluble and particulate fraction obtained by centrifugation at 40,000 × *g* for 10 min. The activity in supernatant and pellet is expressed as a percent of the total activity in the original homogenate. Each point is the mean ± SD for three animals. The values at 40 sec and 10 min were compared for statistically significant differences from the corresponding (soluble or mitochondrial) value at 20 sec; it was technically impossible to assess the soluble-particulate distribution at earlier times of ischemia. Single star indicates $p < 0.1$; double star indicates $p < 0.025$. A rapid decrease in soluble activity, with corresponding increase in particulate activity, was seen after less than a minute of ischemia, with no further changes even after 10 min of ischemia. A somewhat tenuous extrapolation of the curves suggests an approximately 50:50 distribution of the activity between soluble and particulate fraction prior to ischemia. These results were obtained by Y. T. Guh in the author's laboratory.

brain homogenates from chicks,[243] mice,[244] and rats (Figure 3). This redistribution can be attributed to the decreased glucose-6-P[208] and glucose-1,6-bisphosphate,[231] and increased P_i[208] levels that result from ischemia. An increased proportion of the total hexokinase is also found in particulate form when glucose-6-P levels are decreased during insulin-induced hypoglycemia,[245] aglycemic perfusion of the brain,[246] or by a high galactose diet.[243] It was demonstrated that the effects of hypoglycemia[245] and galactose feeding[243] on both glucose-6-P levels and hexokinase distribution could be rapidly and synchronously reversed to normal values simply by intraperitoneal injection of glucose; reversibility by glucose of the changes seen during aglycemic perfusion was apparently not determined.[246] The conditions mentioned thus far represent stresses on brain energy metabolism. In each case, increased particulate hexokinase, attributable to decreased glucose-6-P (and possibly glucose-1,6-bisphosphate also) and increased P_i, was observed. The reverse situation is anesthesia, during which glucose utilization is decreased,[185,186,208] glucose-6-P levels increased, and P_i levels decreased.[208] These changes would be predicted[113] to result in an increased proportion of the total hexokinase in the *soluble* fraction of brain homogenates, and this has been demonstrated[246-251] using a variety of anesthetics. One[251] of these studies included the use of the "brain blower"[212] in an attempt to minimize post-mortem changes in hexokinase distribution or metabolite levels. In contrast to the results of Höfeler and Krieglstein,[251] Kyriazi and Basford,[252] also using a "brain blower", have reported no effect of nembutal anesthesia on distribution of hexokinase; the resolution of this disagreement awaits further work.

Although the increased soluble hexokinase activity observed during anesthesia[246-251] can reasonably be attributed to the increased glucose-6-P and decreased P_i that accompany this state,[208] a direct effect of anesthetics on the outer mitochondrial membrane has also been suggested.[247,248,250,251] Various anesthetics, at concentrations estimated to be in the range of therapeutic use, cause solubilization of hexokinase from isolated brain mitochondria, while other lipophilic drugs do not.[247-251] It may also be noted here that the anesthetic thiopental has been reported to cause increased solubilization of mitochondrial hexokinase in various tumor cells;[253] there were no measurements of the effect, if any, of thiopental on cellular glucose-6-P or P_i content.

When changes in the intracellular distribution of brain hexokinase were first proposed to be of importance in its regulation,[113] this was envisaged as an equilibrium involving only two forms of the enzyme, one bound to the mitochondria and one completely dissociated from that organelle. Subsequent work has indicated that this initial view should be modified to include an additional, more loosely bound form of the enzyme as an intermediate in the equilibrium between a tightly bound mitochondrial enzyme and completely dissociated hexokinase. The most important observations that prompted inclusion of this intermediate were that in the presence of Mg^{++}, the enzyme can maintain its association with the mitochondria[97,113] despite the presence of glucose-6-P which induces conformational changes that result in weakened interactions with the membrane and inhibition.[131,144,224] In other words, interactions between hexokinase and the outer mitochondrial membrane must include two components, only one of which is susceptible to disruption by glucose-6-P. The specific nature of these two components was then deduced from studies which implicated divalent cation-mediated interactions between negatively charged groups on the enzyme and membrane[112] as well as specific interactions with a hexokinase-binding protein,[117] now known to be the pore-forming protein[123,124] of the outer mitochondrial membrane (Section III.C). A current view of the interactions between the outer mitochondrial membrane and brain hexokinase is shown in Figure 4. It should be evident that this inclusion of a more loosely bound intermediate in the equilibrium between tightly bound and fully solubilized enzyme does not affect the essential feature of the original proposal,[113] i.e., that reversible interaction between hexokinase and the outer mitochondrial membrane, governed by metabolites such as glucose-6-P and P_i, is an important factor in regulation of brain hexokinase activity in

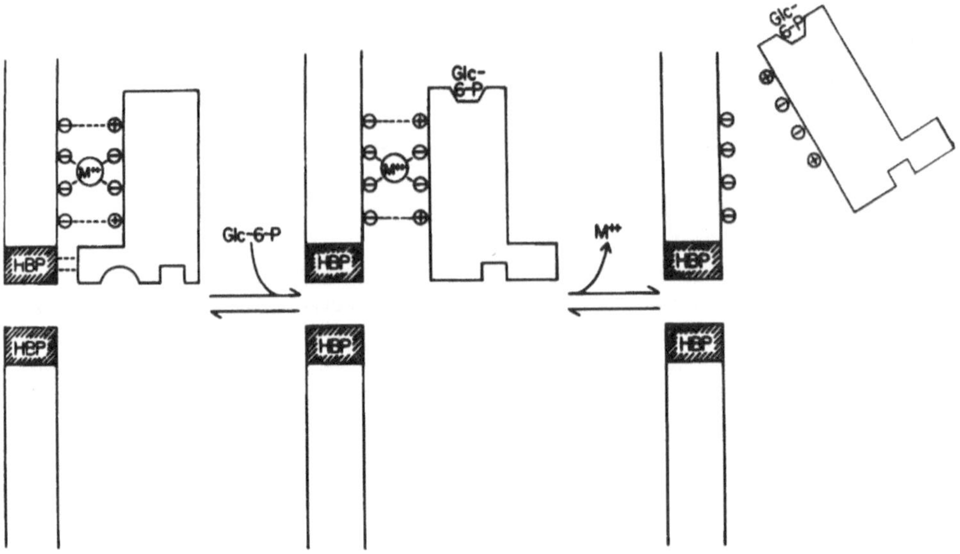

FIGURE 4. Reversible interactions between brain hexokinase and the outer mitochondrial membrane. In the absence of glucose-6-P (left of figure), the enzyme exists in an enzymatically active (i.e., possesses an adenine nucleotide binding site,[132a] see Figure 1) conformation[131,144,224] which interacts with the outer mitochondrial membrane as previously described (legend to Figure 2). Interaction with the hexokinase binding protein may be expected to confer additional stability on the active conformation and thus hinder the conformational change induced by binding of glucose-6-P; it is readily shown[144] that this should decrease the affinity for this ligand, in accord with the observed increase in K_i of the bound enzyme relative to that of the solubilized form.[56,113,178] Binding of glucose-6-P induces a conformational change resulting in inhibition (due to loss of the ability to bind ATP[132a]) and weakened interaction with the membrane,[131,144,224] now maintained only by the electrostatic forces as depicted in the center of the figure, and total dissociation (right of figure) can more readily occur.

vivo. At estimated intracellular levels of Mg^{++} (~1 m*M*),[214] it would be expected that a major portion of the enzyme would remain bound to the mitochondria even in the presence of physiological levels of glucose-6-P.[97,112,113] At one time this was thought to be paradoxical,[12] but further consideration now leads to the realization that retention of a majority of the inactive (due to binding of glucose-6-P) enzyme in association with the mitochondria would be advantageous in that it would facilitate the rapid transition (as glucose-6-P decreases and/or P_i increases) to a fully active mitochondrial enzyme without the need for a possibly difficult recall of substantial amounts of the enzyme from a cytoplasmic (fully solubilized) location. In other words, current thinking suggests that, in vivo, the majority of the enzyme exists in either the active tightly bound form or an inactive loosely bound form (left and center of Figure 4). Since both forms equilibrate with the fully dissociated enzyme (right of Figure 4), this remains consistent with the predicted[113] and observed[243-251] changes in the levels of soluble hexokinase in response to altered energy status.

The possible importance of changes in mitochondrial binding of hexokinase in other tissues is unknown, though there seems to be no *a priori* reason for discounting this possibility. Gumaa and McLean[254] suggested that this might explain the observed kinetics of glucose utilization following addition of glucose to ascites tumor cells (containing both type I and II isozymes), namely, an initial burst of glucose phosphorylation attributed to the activity of mitochondrially bound enzyme and resulting in increased intracellular glucose-6-P, which then led to solubilization of the enzyme and subsequent glucose utilization at a decreased rate due to increased inhibition of the solubilized enzyme by glucose-6-P. Changes in metabolite levels consistent with this suggestion were observed,[255] but no direct evidence for changes in intracellular distribution of hexokinase was presented. Kosow and Rose[136] stated that addition of glucose had no effect on distribution of ascites tumor hexokinase between mitochondrial and soluble forms, but no details were given. Murakami and Ishibashi[62] found increased type II hexokinase in the soluble fraction of ascites cells incubated for 3 hr with glucose when compared with unincubated cells or cells incubated in the absence of glucose; though no metabolite measurements were reported,[62] based on the results of Gumaa and McLean[255] it seems possible that increased glucose-6-P levels were present and thus that the increased soluble hexokinase[62] represented solubilization from a mitochondrial reservoir.

It is evident that the above discussion has focused on the importance of ligand-induced conformational changes in determining the avidity with which hexokinase binds to the mitochondrial membrane. It might be inferred from this that the membrane itself, aside from providing a location for the hexokinase binding protein, is rather neutral in this interaction. In fact, this probably reflects more the state of our knowledge than it does the real state of affairs. Recent work with Dr. Freddie Moller in our laboratory, has shown that the sensitivity of the mitochondrial enzyme to solubilization by glucose-6-P is markedly affected by the phospholipid composition of the mitochondrial membrane.[256] Specifically, lysophospholipids and acidic phopholipids (anionic at neutral pH) greatly increase the solubilization. The mechanism by which the lysolipids exert their effect is not known but it is definitely not simply a general disruption of the membrane due to the well-known detergent action of these compounds. The effect of the anionic phospholipids is readily explained in terms of the current view of the hexokinase-membrane interaction (Figure 4), since increased amounts of these lipids in the membrane would increase the negative charge on the membrane and hence may increase the repulsive force beyond the masking capability of the bridging divalent cations and thereby potentiate the solubilization action of glucose-6-P. These results make it evident that variation in the phospholipid content of the outer mitochondrial membrane may represent another means by which the intracellular distribution of hexokinase is governed.

VI. CONCLUDING COMMENTS

In his recent brief review of vertebrate hexokinase isozymes, Ureta[20] concluded with the observation that the extensive accumulation of data on hexokinases is worthless until we have an answer to the general question: "Why isozymes?" The present chapter is, in part, an attempt to correlate information leading to an answer to that question. The paucity of data on the type III isozyme provides little raw material for such an effort and increasing the awareness of this deficiency may encourage efforts that will remedy it. On the other hand, there is an abundance of data dealing with the type I and II isozymes, such an abundance that it is indeed a challenge to extract basic concepts regarding the differences in physiological function that might be attributed to these isozymes. The readers will likely differ greatly in their opinion as to how sucessfully the author has met this challenge but, in any case, it is hoped that the present effort will be found useful.

We have developed the thesis that the type I and II isozymes are fundamentally different in their physiological roles, being predominantly catabolic and anabolic, respectively. It is worth noting that, despite the many other physical, chemical, and kinetic similarities of these isozymes, their differing response to P_i — an ion of structural simplicity but profound metabolic significance — can lead to major differences in the part that type I and II hexokinase may play in cellular metabolism.

We have also developed the thesis that binding of hexokinase to the mitochondrial membrane has major consequences for the metabolic function of this enzyme and, referring again to Ureta's question "Why isozymes?", have pointed out the differences between the isozymes in this regard. Indeed, the hexokinase isozymes may be an example of what Ureta himself has proposed, namely, that one of the major differences having physiological significance may be in the relative ability of different isozymes to interact specifically with other cellular components.[257] That the interaction with mitochondria may not be a static one, but rather a dynamic equilibrium sensitive to altered cellular needs, introduces an additional aspect of metabolic control, and hexokinase may be but one example of a general class of enzymes exhibiting such behavior.[258]

ACKNOWLEDGMENTS

To the many investigators whose work has furnished the basis for this chapter, I am greatly indebted. Our own work, some of which is cited in the present article, has received continuing support from NIH Grant NS 09910, for which we are most grateful.

REFERENCES

1. **Elbrink, J. and Bihler, I.,** Membrane transport: its relation to cellular metabolic rates, *Science,* 188, 1177, 1975.
2. **Sols, A.,** Phosphorylation and glycolysis, in *Carbohydrate Metabolism and Its Disorders,* Dickens, F., Randle, P. J., and Whelan, W. J., Eds., Academic Press, New York, 1968, chap. 3.
3. **Ramaiah, A.,** Pasteur effect and phosphofructokinase, in *Current Topics in Cellular Regulation,* Vol. 8, Horecker, B. L. and Stadtman, E. R., Eds., Academic Press, New York, 1974, 298.
4. **Vallejo, C. G., Marco, R., and Sebastián, J.,** The glucose-6-phosphate metabolic crossroads in brain. Studies at the enzyme level, *Arch. Biochem. Biophys.,* 147, 41, 1971.
5. **Bonsignore, A. and DeFlora, A.,** Regulatory properties of glucose-6-phosphate dehydrogenase, in *Current Topics in Cellular Regulation,* Vol. 6, Horecker, B. L. and Stadtman, E. R., Eds., Academic Press, New York, 1972, 21.
6. **Sols, A. and Marco, R.,** Concentration of metabolites and binding sites. Implications in metabolic regulation, in *Current Topics in Cellular Regulation,* Vol. 2, Horecker, B. L. and Stadtman, E. R., Eds., Academic Press, New York, 1970, 227.
7. **Masters, C. J.,** Metabolic control and the microenvironment, in *Current Topics in Cellular Regulation,* Vol. 12, Horecker, B. L. and Stadtman, E. R., Eds., Academic Press, New York, 1977, 75.
8. **Ottaway, J. H. and Mowbray, J.,** The role of compartmentation in the control of glycolysis, in *Current Topics in Cellular Regulation,* Vol. 12, Horecker, B. L. and Stadtman, E. R., Eds., Academic Press, New York, 1977, 108.
9. **Srere, P. and Estabrook, R. W., Eds.,** *Microenvironments and Metabolic Compartmentation,* Academic Press, New York, 1978.
10. **Colowick, S. P.,** The hexokinases, in *The Enzymes,* Vol. 9, 3rd ed., Boyer, P. D., Ed., Academic Press, New York, 1973, 1.
11. **Purich, D. L., Fromm, H. J., and Rudolph, F. B.,** The hexokinases: kinetic, physical, and regulatory properties, in *Advances in Enzymology,* Vol. 39, Meister, A., Ed., John Wiley & Sons, New York, 1973, 249.
12. **Wilson, J. E.,** Brain hexokinase, the prototype ambiquitous enzyme, in *Current Topics in Cellular Regulation,* Vol. 16, Horecker, B. L. and Stadtman, E. R., Eds., Academic Press, New York, 1980, 1.
13. **Wilson, J. E.,** Hexokinase, in *Handbook of Neurochemistry,* Vol. 4, 2nd ed., Lajtha, A., Ed., Plenum Press, New York, 1983, 151.
14. **Bessman, S. P. and Geiger, P. J.,** Compartmentation of hexokinase and creatine phosphokinase, cellular regulation, and insulin action, in *Current Topics in Cellular Regulation,* Vol. 16, Horecker, B. L. and Stadtman, E. R., Eds., Academic Press, New York, 1980, 55.
15. **Viñuela, E., Salas, M., and Sols, A.,** Glucokinase and hexokinase in liver in relation to glycogen synthesis, *J. Biol. Chem.,* 238, PC 1175, 1963.
16. **Walker, D. G.,** On the presence of two soluble glucose-phosphorylating enzymes in adult liver and the development of one of these after birth, *Biochim. Biophys. Acta,* 77, 209, 1963.

17. **González, C., Ureta, T., Babul, J., Rabajille, E., and Niemeyer, H.,** Characterization of isozymes of adenosine triphosphate: D-hexose 6-phosphotransferase from rat liver, *Biochemistry*, 6, 460, 1967.
18. **Katzen, H. M., Soderman, D. D., and Nitowsky, H. M.,** Kinetic and electrophoretic evidence for multiple forms of glucose-ATP phosphotransferase activity from human cell cultures and rat liver, *Biochem. Biophys. Res. Commun.*, 19, 377, 1965.
19. **Katzen, H. M. and Schimke, R. T.,** Multiple forms of hexokinase in the rat: tissue distribution, age dependency, and properties, *Proc. Natl. Acad. Sci. U.S.A.*, 54, 1218, 1965.
20. **Ureta, T.,** The comparative isozymology of vertebrate hexokinases, *Comp. Biochem. Physiol.*, 71B, 549, 1982.
21. *Enzyme Nomenclature*, Elsevier, Amsterdam, 1973, 23.
22. **Katzen, H. M.,** The multiple forms of mammalian hexokinase and their significance to the action of insulin, in *Advances in Enzyme Regulation*, Vol. 5, Weber, G., Ed., Pergamon Press, New York, 1967, 335.
23. **Sols, A., Salas, M., and Viñuela, E.,** Induced biosynthesis of liver glucokinase, in *Advances in Enzyme Regulation*, Vol. 2, Weber, G., Ed., Pergamon Press, New York, 1964, 177.
24. **Grossbard, L. and Schimke, R. T.,** Multiple hexokinases of rat tissues: purification and comparison of soluble forms, *J. Biol. Chem.*, 241, 3546, 1966.
25. **Davagnino, J. and Ureta, T.,** The identification of extrahepatic "glucokinase" as N-acetylglucosamine kinase, *J. Biol. Chem.*, 255, 2633, 1980.
26. **Meglasson, M. D., Burch, P. T., Berner, D. K., Najafi, H., Vogin, A. P., and Matschinsky, F. M.,** Chromatographic resolution and kinetic characterization of glucokinase from islets of Langerhans, *Proc. Natl. Acad. Sci. U.S.A.*, 80, 85, 1983.
27. **Sapag-Hagar, M., Marco, R., and Sols, A.,** Distribution of hexokinase and glucokinase between parenchymal and non-parenchymal cells of rat liver, *FEBS Lett.*, 76, 226, 1977.
28. **Crisp, D. M. and Pogson, C. I.,** Glycolytic and gluconeogenic enzyme activities in parenchymal and non-parenchymal cells from mouse liver, *Biochem. J.*, 126, 1009, 1972.
29. **Bontemps, F., Hue, L., and Hers, H.-G.,** Phosphorylation of glucose in isolated rat hepatocytes, *Biochem. J.*, 174, 603, 1978.
30. **Niemeyer, H., Cárdenas, M. D., Rabajille, E., Ureta, T., Clark-Turri, L., and Peñaranda, J.,** Sigmoidal kinetics of glucokinase, *Enzyme*, 20, 321, 1975.
31. **Storer, A. C. and Cornish-Bowden, A.,** Kinetics of rat liver glucokinase, *Biochem. J.*, 159, 7, 1976.
32. **Storer, A. C. and Cornish-Bowden, A.,** Kinetic evidence for a "mnemonical" mechanism for rat liver glucokinase, *Biochem. J.*, 165, 61, 1977.
33. **Ashcroft, S. J. H., Hedeskov, C. J., and Randle, P. J.,** Glucose metabolism in mouse pencreatic islets, *Biochem. J.*, 118, 143, 1970.
34a. **Zawalich, W. S. and Matschinsky, F.,** Analysis of the releasing and fuel function of glucose in isolated perifused pancreatic islets, *Endocrinology*, 100, 1, 1977.
34b. **Shatton, J. B., Morris, H. P., and Weinhouse, S.,** Kinetic, electrophoretic, and chromatographic studies on glucose-ATP- phosphotransferases in rat hepatomas, *Cancer Res.*, 25, 1161, 1969.
34c. **Meglasson, M. D., Burch, P. T., Hoenig, M., Chick, W. L., and Matschinsky, F. M.,** Identification and significance of glucokinase in transplantable insulinomas, *J. Biol. Chem.*, 258, 2094, 1983.
35. **Walker, D. G. and Holland, G.,** The development of hepatic glucokinase in the neonatal rat, *Biochem. J.*, 97, 845, 1965.
36. **Jamdar, S. C. and Greengard, O.,** Premature formation of glucokinase in developing rat liver, *J. Biol. Chem.*, 245, 2779, 1970.
37. **Ureta, T., González, C., Lillo, S., and Niemeyer, H.,** Comparative studies on glucose phosphorylating isozymes of vertebrates. I. The influence of fasting and the nature of the diet on liver glucokinase and hexokinases of rodents, *Comp. Biochem. Physiol.*, 40B, 71, 1971.
38. **Povey, S., Corney, G., and Harris, H.,** Genetically determined polymorphism of hexokinase, HK III, found in human leucocytes, *Ann. Hum. Genet. London*, 38, 407, 1975.
39. **Ritter, H., Friedrickson, U., and Schmitt, J.,** Genetic polymorphism of hexokinase in primates, *Humangenetik*, 22, 265, 1974.
40. **Rogers, P. A., Fisher, R. A., and Harris, H.,** An electrophoretic study of the distribution and properties of human hexokinases, *Biochem. Genet.*, 13, 857, 1975.
41. **Ueda, M., Taketa, K., and Kosaka, K.,** Hexokinase isozyme pattern in CCl_4-injured rat liver, *Clin. Chim. Acta*, 60, 77, 1975.
42. **Felgner, P. L. and Wilson, J. E.,** Purification of nonbindable and membrane-bindable mitochondrial hexokinase from rat brain, *Biochem. Biophys. Res. Commun.*, 68, 592, 1976.
43. **Needels, D. L. and Wilson, J. E.,** The identity of hexokinase activities from mitochondrial and cytoplasmic fractions of rat brain homogenates, *J. Neurochem.*, 40, 1134, 1983.
44. **Vowles, D. T. and Easterby, J. S.,** Comparison of type I hexokinases from pig heart and kinetic evaluation of the effects of inhibitors, *Biochim. Biophys. Acta*, 566, 283, 1979.

45. **Gellerich, F. N. and Augustin, H. W.**, Electrophoretic characterization and subcellular distribution of hexokinase isozymes in red blood cells of rabbits, *Acta Biol. Med. Germ.*, 38, 1091, 1979.
46. **Rijksen, G., Jansen, G., Kraaijenhagen, R. J., Van der Vlist, M. J. M., Vlug, A. M. C., and Staal, G. E. J.**, Separation and characterization of hexokinase I subtypes from human erythrocytes, *Biochim. Biophys. Acta*, 659, 292, 1981.
47. **Stocchi, V., Magnani, M., Canestrari, F., Dacha, M., and Fornaini, G.**, Rabbit red blood cell hexokinase — evidence for two distinct forms, and their purification and characterization from reticulocytes, *J. Biol. Chem.*, 256, 7856, 1981.
48. **Stocchi, V., Magnani, M., Canestrari, F., Dacha, M., and Fornaini, G.**, Multiple forms of human red blood cell hexokinase — purification, characterization and age dependence, *J. Biol. Chem.*, 257, 2357, 1982.
49. **Neumann, S., Falkenburg, F., and Pfleiderer, G.**, Purification and immunological characterization of the human hexokinase isozymes I and III (ATP-D-hexose 6-phosphotransferase EC 2.7.1.1), *Biochim. Biophys. Acta*, 334, 328, 1974.
50. **Faulkner, A. and Jones, C. T.**, Hexokinase isoenzymes in tissues of the adult and developing guinea pig, *Arch. Biochem. Biophys.*, 175, 477, 1976.
51. **Katzen, H. M. and Soderman, D. D.**, The hexokinase isozymes: sulfhydryl considerations in the regulation of the particle-bound and soluble states, in *Isozymes*, Vol. 2, Markert, C., Ed., Academic Press, New York, 1975, 797.
52. **Katzen, H. M., Soderman, D. D., and Wiley, C. E.**, Multiple forms of hexokinase, *J. Biol. Chem.*, 245, 4081, 1970.
53a. **Grossbard, L., Weksler, M., and Schimke, R. T.**, Electrophoretic properties and tissue distribution of multiple forms of hexokinase in various mammalian species, *Biochem. Biophys. Res. Commun.*, 24, 32, 1966.
53b. **Magnani, M., Stocchi, V., Canestrari, F., Dacha, M., and Fornaini, G.**, Multiple forms of hexokinase in mammalian red blood cells, *Biochem. Int.*, 4, 673, 1982.
54. **Neumann, S. and Pfleiderer, G.**, Immunological specificity of the isozymes I and III of human hexokinase (ATP: D-hexose 6-phosphotransferase EC 2.7.1.1), *Biochim. Biophys. Acta*, 334, 343, 1974.
55. **McLean, P., Brown, J., Walters, E., and Greenslade, K.**, Effect of alloxan-diabetes on multiple forms of hexokinase in adipose tissue and lung, *Biochem. J.*, 105, 1301, 1967.
56. **Salotra, P. T. and Singh, V. N.**, Regulation of glucose metabolism in rat lung: subcellular distribution, isozyme pattern, and kinetic properties of hexokinase, *Arch. Biochem. Biophys.*, 216, 758, 1982.
57. **Ureta, T., Radojkoviac, J., Zepeda, S., and Guixe, V.**, Comparative studies on glucose phosphorylating isoenzymes of vertebrates. VII. Mammalian hexokinases, *Comp. Biochem. Physiol.*, 70B, 225, 1981.
58. **Knull, H. R. and Lamb, S. M.**, Hexokinase in salivary glands, *Comp. Biochem. Physiol.*, 65B, 743, 1980.
59. **Dean, J. C. and Gusseck, D. J.**, Changing patterns of placental hexokinase isozymes during the course of gestation, *Arch. Biochem. Biophys.*, 172, 130, 1976.
60. **Kraaijenhagen, R. J., Rijsken, G., and Staal, G. E. J.**, Hexokinase isozyme distribution and regulatory properties in lymphoid cells, *Biochim. Biophys. Acta*, 631, 402, 1980.
61. **Rijsken, G., Staal, G. E. J., Beks, P. J., Streefkerk, M., and Akkerman, J. W. N.**, Compartmentation of hexokinase in human blood cells — characterization of soluble and particulate enzymes, *Biochim. Biophys. Acta*, 719, 431, 1982.
62. **Murakami, K. and Ishibashi, S.**, Variation of hexokinase isoenzyme pattern in Ehrlich ascites tumor cells in relation to the utilization of glucose in the cells, *J. Biochem. (Tokyo)*, 73, 207, 1973.
63. **Katzen, H. M.**, Multiple forms of hexokinase in the rat: starch gel electrophoretic evidence for an effect of diabetes and insulin *in vitro*, *Diabetes*, 15, 518, 1966.
64. **Rose, I. A. and Warms, J. V. B.**, Stability of hexokinase II *in vitro* and in ascites tumor cells, *Arch. Biochem. Biophys.*, 213, 625, 1982.
65. **Brewer, G. J.**, General red cell metabolism, in *The Red Blood Cell*, Vol. 1, 2nd ed., Surgenor, D. M., Ed., Academic Press, New York, 1974, chap. 9.
66. **Ballatori, N. and Cohen, J. J.**, Intracellular distribution of hexokinase in the tissue zones of rat kidney, *Biochim. Biophys. Acta*, 657, 448, 1981.
67. **Cahill, G. F., Jr. and Owen, O. E.**, Some observations on carbohydrate metabolism in man, in *Carbohydrate Metabolism and Its Disorders*, Dickens, F., Randle, P. J., and Whelan, W. J., Eds., Academic Press, New York, 1968, chap. 16.
68. **Krebs, H. A.**, Some aspects of the regulation of fuel supply in omnivorous animals, in *Advances in Enzyme Regulation*, Vol. 10, Weber, G., Ed., Pergamon Press, Elmsford, N.Y., 1972, 397.
69. **Morgan, H. E., Henderson, M. J., Regen, D. M., and Park, C. R.**, Regulation of glucose uptake in muscle, *J. Biol. Chem.*, 236, 253, 1961.
70. **Wilson, J. E. and Felgner, P. L.**, An inverse relation between mitochondrial hexokinase content and phosphoglucomutase activity of rat tissues, *Mol. Cell. Biochem.*, 18, 39, 1977.

71. **Sherman, T. G. and Wilson, J. E.,** Effect of prenatally-induced and postnatally-maintained ketosis on β-hydroxybutyrate dehydrogenase and hexokinase levels in the developing rat brain, *J. Neurochem.*, 30, 639, 1978.
72. **Walters, E. and McLean, P.,** Multiple forms of glucose-adenosine triphosphate phosphotransferase in rat mammary gland, *Biochem. J.*, 104, 778, 1967.
73. **Walters, E. and McLean, P.,** The effect of anti-insulin serum and alloxan-diabetes on the distribution and multiple forms of hexokinase in lactating rat mammary gland, *Biochem. J.*, 109, 737, 1968.
74. **Schwartz, G. P. and Basford, R. E.,** The isolation and purification of solubilized hexokinase from bovine brain, *Biochemistry*, 6, 1070, 1967.
75. **Redkar, V. D. and Kenkare, U. W.,** Bovine brain mitochondrial hexokinase, *J. Biol. Chem.*, 247, 7576, 1972.
76. **Chou, A. C. and Wilson, J. E.,** Purification and properties of rat brain hexokinase, *Arch. Biochem. Biophys.*, 151, 48, 1972.
77. **Magnani, M., Serafini, G., Stocchi, V., Bossu, M., and Dacha, M.,** Solubilization, purification, and properties of rabbit brain hexokinase, *Arch. Biochem. Biophys.*, 216, 449, 1982.
78. **Easterby, J. S. and O'Brien, M. J.,** Purification and properties of pig heart hexokinase, *Eur. J. Biochem.*, 38, 201, 1973.
79. **Farmer, E. E. and Easterby, J. S.,** Purification of heart hexokinase by dye-ligand chromatography, *Anal. Biochem.*, 123, 373, 1982.
80. **Wright, C. L., Warsy, A. S., Holroyde, M. J., and Trayer, I. P.,** Purification of the hexokinases by affinity chromatography on Sepharose-N-aminoacylglucosamine derivatives, *Biochem. J.*, 175, 125, 1978.
81. **Magnani, M., Dacha, M., Stocchi, V., Ninfali, P., and Fornaini, G.,** Rabbit red blood cell hexokinase-purification and properties, *J. Biol. Chem.*, 255, 1752, 1980.
82. **Holroyde, M. J. and Trayer, I. P.,** Purification and properties of rat skeletal muscle hexokinase, *FEBS Lett.*, 62, 215, 1976.
83. **Qadri, S. S. and Easterby, J. S.,** Purification of skeletal muscle hexokinase by affinity elution chromatography, *Anal. Biochem.*, 105, 299, 1980.
84. **Holroyde, M. J., Allen, M. B., Storer, A. C., Warsy, A. S., Chesher, J. M., Trayer, I. P., Cornish-Bowden, A., and Walker, D. G.,** The purification in high yield and characterization of rat hepatic glucokinase, *Biochem. J.*, 153, 363, 1976.
85. **Grossman, S. H., Dorn, C. G., and Potter, V. R.,** The preparation and characterization of pure rat liver glucokinase, *J. Biol. Chem.*, 249, 3055, 1974.
86. **Maccioni, R. and Babul, J.,** Purification and characterization of dog liver glucokinase, *Arch. Biol. Med. Exp.*, 13, 271, 1980.

87a. **Cornish-Bowden, A.,** Critical values for testing the significance of amino acid composition indexes, *Anal. Biochem.*, 105, 233, 1980.

87b. **Darby, M. K. and Trayer, I. P.,** Evolutionary relationships among the hexokinases by thin-layer peptide mapping, *Biochem. Soc. Trans.*, 9, 62, 1981.

88. **Pilkis, S. J., Hansen, R. J., and Krahl, M. E.,** Hexose-ATP phosphotransferases: comparative aspects, *Comp. Biochem. Physiol.*, 25, 903, 1968.
89. **Clark-Turri, L., Peñaranda, J., Rabajille, E., and Niemeyer, H.,** Immunochemical titration of liver glucokinase from normal, fasted, and diabetic rats, *FEBS Lett.*, 41, 342, 1974.
90. **Creighton, S. R., McClure, A. M., Watrous, B. J., and Hansen, R. J.,** Hexose-ATP phosphotransferases: comparative aspects. III. Interrelationships of animal hexokinases, *Comp. Biochem. Physiol.*, 42B, 509, 1972.
91. **Craven, P. A. and Basford, R. E.,** Brain hexokinase: immunohistochemical localization at the light microscopic level, *Biochemistry*, 8, 3520, 1969.
92. **Ureta, T., Radojković, J., Lagos, R., Guixé, V., and Nuñez, L.,** Phylogenetic and ontogenetic studies of glucose phosphorylating isozymes of vertebrates, *Arch. Biol. Med. Exp.*, 12, 587, 1979.
93. **Atassi, M. Z. and Lee, C-L.,** Boundary refinement of the lysozyme antigenic site around the disulphide bond 6-127 (site 1) by "surface-simulation" synthesis, *Biochem. J.*, 171, 419, 1978.
94. **Lerner, R. A.,** Tapping the immunological repertoire to produce antibodies of predetermined specificity, *Nature (London)*, 299, 592, 1982.
95. **Crane, R. K. and Sols, A.,** The association of hexokinase with particulate fractions of brain and other tissue homogenates, *J. Biol. Chem.*, 203, 273, 1953.
96. **Johnson, M. K.,** The intracellular distribution of glycolytic and other enzymes in rat-brain homogenates and mitochondrial preparations, *Biochem. J.*, 77, 610, 1960.
97. **Rose, I. A. and Warms, J. V. B.,** Mitochondrial hexokinase: release, rebinding, and location, *J. Biol. Chem.*, 242, 1635, 1967.
98. **Bachelard, H. S.,** The subcellular distribution and properties of hexokinases in the guinea-pig cerebral cortex, *Biochem. J.*, 104, 286, 1967.

99. **Wilson, J. E.,** The localization of latent brain hexokinase on synaptosomal mitochondria, *Arch. Biochem. Biophys.*, 150, 96, 1972.
100. **Hernandez, A. and Crane, R. K.,** Association of heart hexokinase with subcellular structure, *Arch. Biochem. Biophys.*, 113, 223, 1966.
101. **Mayer, S. E., Mayfield, A. C., and Haas, J. A.,** Heart muscle hexokinase: subcellular distribution and inhibition by glucose 6-phosphate, *Mol. Pharmacol.*, 2, 393, 1966.
102. **Srivastava, L. M., Shakespeare, P., and Hübscher, G.,** Glucose metabolism in the mucosa of the small intestine, *Biochem. J.*, 109, 35, 1968.
103. **Anderson, J. W. and King, P.,** Subcellular distribution of hexokinase activity in rat jejunal mucosa: response to diabetes and dietary changes, *Biochem. Med.*, 12, 1, 1975.
104. **Hochachka, P. W., Liggins, G. C., Qvist, J., Schneider, R., Snider, M. Y., Wonders, T. R., and Zapol, W. M.,** Pulmonary metabolism during diving: conditioning blood for the brain, *Science*, 198, 831, 1977.
105. **Borrebaek, B.,** Mitochondrial-bound hexokinase of the rat epididymal adipose tissue and its possible relation to the action of insulin, *Biochem. Med.*, 3, 485, 1970.
106. **Toews, C. J.,** Intracellular distribution and some properties of hexokinase of bovine adrenal medulla, *Can. J. Biochem.*, 44, 879, 1966.
107. **Southard, J. H. and Hultin, H. O.,** On latent hexokinase activity in skeletal muscle mitochondria, *FEBS Lett.*, 19, 349, 1972.
108. **Baquer, N. A. and McLean, P.,** The effect of oestrogen on the activity and binding of multiple forms of hexokinase in the rat uterus, *Biochem. Biophys. Res. Commun.*, 37, 158, 1969.
109. **Bartley, J. C., Barber, S., and Abraham, S.,** Identity, release, and binding of mitochondrial-bound hexokinases in mammary glands and adenocarcinomas of lactating mice, *Cancer Res.*, 35, 1649, 1975.
110. **Etinghof, R. N., Biesold, D., Teichgräber, P., and Müller, L.,** Mitochondrial hexokinase of retina: distribution, activation, and elution, *Biokhimya*, 34, 1034, 1969.
111. **Kurokawa, M., Oda, S., Tsubotani, E., Fujiwara, H., Yokoyama, K., and Ishibashi, S.,** Characterization of hexokinase isoenzyme Types I and II in ascites tumor cells by an interaction with mitochondrial membrane, *Mol. Cell. Biochem.*, 45, 151, 1982.
112. **Felgner, P. L. and Wilson, J. E.,** Effect of neutral salts on the interaction of rat brain hexokinase with the outer mitochondrial membrane, *Arch. Biochem. Biophys.*, 182, 282, 1977.
113. **Wilson, J. E.,** Brain hexokinase — a proposed relation between soluble-particulate distribution and activity, *in vivo*, *J. Biol. Chem.*, 243, 3640, 1968.
114. **Teichgräber, P. and Biesold, D.,** Properties of membrane-bound hexokinase in rat brain, *J. Neurochem.*, 15, 979, 1968.
115. **Craven, P. A., Goldblatt, P. J., and Basford, R. E.,** Brain hexokinase. The preparation of inner and outer mitochondrial membranes, *Biochemistry*, 8, 3525, 1969.
116. **Kropp, E. S. and Wilson, J. E.,** Hexokinase binding sites on mitochondrial membranes, *Biochem. Biophys. Res. Commun.*, 38, 74, 1970.
117. **Felgner, P. L., Messer, J. L., and Wilson, J. E.,** Purification of a hexokinase-binding protein from the outer mitochondrial membrane, *J. Biol. Chem.*, 254, 4946, 1979.
118. **Font, B., Vial, C., and Gautheron, D. C.,** Intracellular and submitochondrial localization of pig heart hexokinase, *FEBS Lett.*, 56, 24, 1975.
119. **Mayer, R. J. and Hübscher, G.,** Mitochondrial hexokinase from small-intestinal mucosa and brain, *Biochem. J.*, 124, 491, 1971.
120. **Vallejo, C. G., Marco, R., and Sebastián, J.,** The association of brain hexokinase with mitochondrial membranes, *Eur. J. Biochem.*, 14, 478, 1970.
121. **Mayer, R. J.,** Brain mitochondrial hexokinase and adenylate kinase: effect of osmotic conditions on enzyme activities, *J. Neurochem.*, 19, 2127, 1972.
122. **Schlame, M., Gellerich, F. N., and Augustin, W.,** Localization of hexokinase in mitochondria from rabbit reticulocytes and its relation to mitochondrial ATP-formation studied by measurement of ^{32}P-fluxes, *Acta Biol. Med. Germ.*, 40, 617, 1981.
123. **Linden, M., Gellerfors, P., and Nelson, B. D.,** Pore protein and the hexokinase-binding protein from the outer membrane of rat liver mitochondria are identical, *FEBS Lett.*, 141, 189, 1982.
124. **Fiek, C., Benz, R., Roos, N., and Brdiczka, D.,** Evidence for identity between the hexokinase-binding protein and the mitochondrial porin in the outer membrane of rat liver mitochondria, *Biochim. Biophys. Acta*, 688, 429, 1982.
125. **Fromm, H. J.,** Mechanism and mode of regulation of brain hexokinase, in *The Regulation of Carbohydrate Formation and Utilization in Mammals*, Veneziale, C. M., Ed., University Park Press, Baltimore, 1981, 45.
126. **Lueck, J. D. and Fromm, H. J.,** Kinetics, mechanism, and regulation of rat skeletal muscle hexokinase, *J. Biol. Chem.*, 249, 1341, 1974.

127. **Cayanis, E. and Balinsky, D.**, Comparative kinetic properties of human hexokinases, *Int. J. Biochem.*, 6, 741, 1975.
128. **Segel, I. H.**, *Enzyme Kinetics*, John Wiley & Sons, New York, 1975, chap. 9.
129. **Bachelard, H. S., Clark, A. G., and Thompson, M. F.**, Cerebral cortex hexokinases, *Biochem. J.*, 123, 707, 1971.
130. **Gerber, G., Preissler, H., Heinrich, R., and Rapoport, S. M.**, Hexokinase of human erythrocytes, *Eur. J. Biochem.*, 45, 39, 1974.
131. **Wilson, J. E.**, Ligand-induced conformations of rat brain hexokinase: studies with hexoses, hexose 6-phosphates, and nucleoside triphosphates leading to development of a new model for the enzyme, *Arch. Biochem. Biophys.*, 185, 88, 1978.
132a. **Baijal, M. and Wilson, J. E.**, Studies on the binding of nucleotides by rat brain hexokinase, *Arch. Biochem. Biophys.*, 218, 513, 1982.
132b. **Gregoriou, M., Cornish-Bowden, A., and Trayer, I. P.**, Isotope exchange evidence for allosteric regulation of hexokinase II by glucose 6-phosphate and for an obligatory addition of substrates, *Biochem. Soc. Trans.*, 9, 62, 1981.
133. **Tuttle, J. P. and Wilson, J. E.**, Rat brain hexokinase: a kinetic comparison of soluble and particulate forms, *Biochim. Biophys. Acta*, 212, 185, 1970.
134. **Kosow, D. P. and Rose, I. A.**, Product inhibition of the hexokinases, *J. Biol. Chem.*, 245, 198, 1970.
135. **Ning, J., Purich, D. L., and Fromm, H. J.**, Studies on the kinetic mechanism and allosteric nature of bovine brain hexokinase, *J. Biol. Chem.*, 244, 3840, 1969.
136. **Kosow, D. P. and Rose, I. A.**, Origin of the delayed feedback control of glucose utilization in ascites tumor cells, *Biochem. Biophys. Res. Commun.*, 48, 376, 1972.
137. **Kosow, D. P., Oski, F. A., Warms, J. V. B., and Rose, I. A.**, Regulation of mammalian hexokinase: regulatory differences between isoenzyme I and II, *Arch. Biochem. Biophys.*, 157, 114, 1973.
138. **Crane, R. K. and Sols, A.**, The non-competitive inhibition of brain hexokinase by glucose-6-phosphate and related compounds, *J. Biol. Chem.*, 210, 597, 1954.
139. **Casazza, J. P. and Fromm, H. J.**, Studies on the mode of sugar-phosphate product inhibition of brain hexokinase, *Arch. Biochem. Biophys.*, 177, 480, 1976.
140. **Lazo, P. A., Sols, A., and Wilson, J. E.**, Brain hexokinase has two spatially discrete sites for binding of glucose 6-phosphate, *J. Biol. Chem.*, 255, 7548, 1980.
141. **Solheim, L. P. and Fromm, H. J.**, Kinetic evidence that the high-affinity glucose 6-phosphate site on hexokinase I is the active site, *Arch. Biochem. Biophys.*, 211, 92, 1981.
142. **Ellison, W. R., Lueck, J. D., and Fromm, H. J.**, Studies on the kinetics and mechanism of orthophosphate activation of bovine brain hexokinase, *Biochem. Biophys. Res. Commun.*, 57, 1214, 1974.
143. **Ellison, W. R., Lueck, J. D., and Fromm, H. J.**, Studies on the mechanism of orthophosphate regulation of bovine brain hexokinase, *J. Biol. Chem.*, 250, 1864, 1975.
144. **Wilson, J. E.**, Ligand-induced conformational changes of rat brain hexokinase: their role in determining the substrate specificity for hexoses and inhibitory effectiveness of hexose 6-phosphates, *Arch. Biochem. Biophys.*, 196, 79, 1979.
145. **Rose, I. A., Warms, J. V. B., and Kosow, D. P.**, Specificity for the glucose-6-P inhibition site of hexokinase, *Arch. Biochem. Biophys.*, 164, 729, 1974.
146. **Rose, I. A. and Warms, J. V. B.**, pH dependence of the α-glucose 1,6-diphosphate inhibition of hexokinase II, *Arch. Biochem. Biophys.*, 171, 678, 1975.
147. **Rijsken, G. and Staal, G. E. J.**, Regulation of human erythrocyte hexokinase by glucose-1,6-diphosphate and inorganic phosphate, *FEBS Lett.*, 80, 61, 1977.
148. **Beitner, R., Haberman, S., and Livni, L.**, Complementarity in the regulation of phosphoglucomutase, phosphofructokinase, and hexokinase; the role of glucose 1,6-bisphosphate, *Biochim. Biophys. Acta*, 397, 355, 1975.
149. **Beitner, R.**, The role of glucose 1,6-bisphosphate in the regulation of carbohydrate metabolism in muscle, *Trends Biochem. Sci.*, 4, 228, 1979.
150. **Wilson, J. E.**, Studies on the molecular weight and lipoprotein nature of glucose-6-phosphate-solubilized rat brain hexokinase, *Arch. Biochem. Biophys.*, 154, 332, 1973.
151. **Rijsken, G. and Staal, G. E. J.**, Regulation of human erythrocyte hexokinase: the influence of glycolytic intermediates and inorganic phosphate, *Biochim. Biophys. Acta*, 485, 75, 1977.
152. **Shone, C. C. and Fromm, H. J.**, Initial rate and inhibition studies on pig brain hexokinase, *Comp. Biochem. Physiol.*, 67B, 697, 1980.
153a. **Rijsken, G. and Staal, G. E. J.**, Human erythrocyte hexokinase deficiency, *J. Clin. Invest.*, 62, 294, 1978.
153b. **Easterby, J. S. and Qadri, S. S.**, Effects of metabolites on the activity of mammalian hexokinases, *Biochem. Soc. Trans.*, 9, 25, 1981.
154. **Sols, A. and Crane, R. K.**, Substrate specificity of brain hexokinase, *J. Biol. Chem.*, 210, 581, 1954.

155. **Bertoni, J. M.,** Competitive inhibition of rat brain hexokinase by 2-deoxyglucose, glucosamine, and metrizamide, *J. Neurochem.*, 37, 1523, 1981.
156. **Bertoni, J. M.,** Metrizamide inhibits human brain hexokinase, *Neurology*, 32, 884, 1982.
157. **Okuda, J., Miwa, I., Inagaki, K., Ueda, M., and Taketa, K.,** D-Glucose anomeric preference of hexokinases in higher animals, *J. Biochem. (Tokyo)*, 84, 993, 1978.
158. **Darby, M. K. and Trayer, I. P.,** Metal-nucleotide structure at the active sites of mammalian hexokinases, *Eur. J. Biochem.*, 129, 555, 1983.
159. **Swarup, G. and Kenkare, U. W.,** Reaction of brain hexokinase with a substrate-like reagent. Alkylation of a single thiol at the active site, *Biochemistry*, 19, 4058, 1980.
160. **Connolly, B. A. and Trayer, I. P.,** Affinity labelling of rat-muscle hexokinase type II by a glucose-derived alkylating agent, *Eur. J. Biochem.*, 93, 375, 1979.
161. **Darby, M., Connolly, B. A., and Trayer, I. P.,** Affinity labelling of the mammalian hexokinases with glucose-derived alkylating agents, *Arch. Biol. Med. Exp.*, 14, 163, 1981.
162. **Subbarao, B. and Kenkare, U. W.,** Inactivation of brain hexokinase by an adenosine 5′-triphosphate analog, *Arch. Biochem. Biophys.*, 181, 19, 1977.
163. **Redkar, V. D. and Kenkare, U. W.,** Effect of ligands on the reactivity of essential sulphydryls in brain hexokinase. Possible interaction between substrate binding sites, *Biochemistry*, 14, 4704, 1975.
164. **Chou, A. C. and Wilson, J. E.,** A study of the sulfhydryl groups of rat brain hexokinase, *Arch. Biochem. Biophys.*, 163, 191, 1974.
165. **Magnani, M., Stocchi, V., Ninfali, P., Dacha, M., and Fornaini, G.,** Action of oxidized and reduced glutathione on rabbit red blood cell hexokinase, *Biochim. Biophys. Acta*, 615, 113, 1980.
166. **Solheim, L. P. and Fromm, H. J.,** pH kinetic studies of bovine brain hexokinase, *Biochemistry*, 19, 6074, 1980.
167. **Wilson, J. E.,** Involvement of arginine residues in catalysis by rat brain hexokinase, *Arch. Biochem. Biophys.*, 218, 254, 1982.
168. **Anderson, C. M., Zucker, F. H., and Steitz, T. A.,** Space-filling models of kinase clefts and conformation changes, *Science*, 204, 375, 1979.
169. **Moore, C. L. and Jöbsis, F. F.,** Some studies on the control of respiration in rat brain mitochondrial preparations, *Arch. Biochem. Biophys.*, 138, 295, 1970.
170. **Siekevitz, P.,** On the meaning of intracellular structure for metabolic regulation, in *Ciba Foundation Symposium on Regulation of Cell Metabolism*, Wolstenholme, G. E. W. and O'Connor, C. M., Eds., Little, Brown, Boston, 1959, 17.
171. **Felgner, P. L.,** Studies on the physiological raison d'être of mitochondrial hexokinase, *Fed. Proc. Fed. Am. Soc. Exp. Biol.*, 32, 488 Abs, 1973.
172. **Gots, R. E. and Bessman, S. P.,** The functional compartmentation of mitochondrial hexokinase, *Arch. Biochem. Biophys.*, 163, 7, 1974.
173. **Bessman, S. P., Borrebaek, B., Geiger, P. J., and Ben-Or, S.,** Mitochondrial creatine kinase and hexokinase: two examples of compartmentation predicted by the hexokinase mitochondrial binding theory of insulin action, in *Microenvironments and Metabolic Compartmentation*, Srere, P. A. and Estabrook, R. W., Eds., Academic Press, New York, 1978, 111.
174. **Viitanen, P. and Geiger, P.,** Heart mitochondrially bound hexokinase: direct demonstration of compartmentation with the use of newly synthesized ATP, *Fed. Proc. Fed. Am. Soc. Exp. Biol.*, 38, 561, 1979.
175. **Kurokawa, M., Tokuoka, S., Oda, S., Tsubotani, E., and Ishibashi, S.,** Difference in efficiency of function between mitochondria-bound hexokinase and non-bound one, *Biochem. Int.*, 2, 645, 1981.
176. **Sauer, L. A.,** A Crabtree-like effect with isolated ascites tumor mitochondria, *Biochem. Biophys. Res. Commun.*, 17, 294, 1964.
177. **Bustamante, E. and Pedersen, P. L.,** High aerobic glycolysis of rat hepatoma cells in culture: role of mitochondrial hexokinase, *Proc. Natl. Acad. Sci. U.S.A.*, 74, 3735, 1977.
178. **Kosow, D. P. and Rose, I. A.,** Ascites tumor mitochondrial hexokinase II, *J. Biol. Chem.*, 243, 3623, 1968.
179. **Erecińska, M. and Wilson, D. F.,** Regulation of cellular energy metabolism, *J. Membr. Biol.*, 70, 1, 1982.
180. **Villar-Palasi, C. and Larner, J.,** Levels of activity of the enzymes of the glycogen cycle in rat tissues, *Arch. Biochem. Biophys.*, 86, 270, 1960.
181. **Whittaker, V. P.,** The synaptosome, in *Handbook of Neurochemistry*, Vol. 2, Lajtha, A., Ed., Plenum Press, New York, 1969, chap. 14.
182. **Balázs, R., Patel, A. J., and Richter, D.,** Metabolic compartments in the brain: their properties and relation to morphological structures, in *Metabolic Compartmentation in the Brain*, Balázs, R. and Cremer, J. E., Eds., John Wiley & Sons, New York, 1972, chap. 13.

183. **Schwartz, W. J., Smith, C. B., Davidsen, L., Savaki, H., Sokoloff, L., Mata, M., Fink, D. J., and Gainer, H.,** Metabolic mapping of functional activity in the hypothalamo-neurohypophysial system of the rat, *Science,* 205, 723, 1979.
184. **Mata, M., Fink, D. J., Gainer, H., Smith, C. B., Davidsen, L., Savaki, H., Schwartz, W. J., and Sokoloff, L.,** Activity-dependent energy metabolism in rat posterior pituitary primarily reflects sodium pump activity, *J. Neurochem.,* 34, 213, 1980.
185. **Sokoloff, L.,** Relation between physiological function and energy metabolism in the central nervous system, *J. Neurochem.,* 29, 13, 1977.
186. **Sokoloff, L.,** Localization of functional activity in the central nervous system by measurement of glucose utilization with radioactive deoxyglucose, *J. Cereb. Blood Flow Metab.,* 1, 7, 1981.
187. **McLean, P.,** Carbohydrate metabolism of mammary tissue. I. Pathways of glucose catabolism in the mammary gland, *Biochim. Biophys. Acta,* 30, 303, 1958.
188. **Kopelovich, L., Abraham, S., McGrath, H., DeOme, K. B., and Chaikoff, I. L.,** Metabolic characteristics of a naturally occurring preneoplastic tissue I. Glycolytic enzyme activities of hyperplastic alveolar nodule outgrowths and adenocarcinomas of mouse mammary gland, *Cancer Res.,* 26, 1534, 1966.
189. **Walker, D. G.,** Developmental aspects of carbohydrate metabolism, in *Carbohydrate Metabolism and its Disorders,* Dickens, F., Randle, P. J., and Whelan, W. J., Eds., Academic Press, New York, 1968, chap. 15.
190. **VanRossum, G. D. V.,** Respiration and glycolysis in liver slices prepared from rats of different foetal and post-natal ages, *Biochim. Biophys. Acta,* 74, 15, 1963.
191. **Ureta, T., Bravo, R., and Babul, J.,** Rat liver hexokinase during development, *Enzyme,* 20, 334, 1976.
192. **Hommes, F. A. and Everts, R. S.,** Particulate and free hexokinase in fetal rat liver, *Biol. Neonate,* 33, 193, 1978.
193. **Warburg, O.,** *Metabolism of Tumors,* Constable, London, 1930.
194. **Racker, E.,** Bioenergetics and the problem of tumor growth, *Am. Sci.,* 60, 56, 1972.
195. **Racker, E. and Spector, M.,** Warburg, effect revisited: merger of biochemistry and molecular biology, *Science,* 213, 303, 1981.
196. **Pedersen, P. L.,** Tumor mitochondria and the bioenergetics of cancer cells, *Prog. Exp. Tumor Res.,* 22, 190, 1978.
197. **Pederson, P. L. and Bustamante, E.,** Hexokinase and the abnormal metabolism of hepatoma cells, in *Heart Creatine Kinase. The Integration of Isozymes for Energy Distribution,* Jacobus, W. E. and Ingwall, J. S., Eds., Williams & Wilkins, Baltimore, 1980, chap. 13.
198. **Bustamante, E., Morris, H. P., and Pedersen, P. L.,** Energy metabolism of tumor cells. Requirement for a form of hexokinase with a propensity for mitochondrial binding, *J. Biol. Chem.,* 256, 8699, 1981.
199. **Sydow, G.,** Über interschiede in der activität and intrazellulären lokalisation der hexokinase zwischen normaler rattenleber and hepatomen, *Acta Biol. Med. Germ.,* 20, 1, 1968.
200. **Saito, M. and Sato, S.,** Studies on particle-bound hexokinase in rat ascites hepatoma cells, *Biochim. Biophys. Acta,* 227, 344, 1971.
201. **Singh, V. N., Singh, M., August, J. T., and Horecker, B. L.,** Alterations in glucose metabolism in chick-embryo cells transformed by Rous sarcoma virus: intracellular levels of glycolytic intermediates, *Proc. Natl. Acad. Sci. U.S.A.,* 71, 4129, 1974.
202. **Singh, M., Singh, V. N., August, J. T., and Horecker, B. L.,** Alterations in glucose metabolism in chick embryo cells transformed by Rous sarcoma virus, *Arch. Biochem. Biophys.,* 165, 240, 1974.
203. **Gobeev, V. N., Davydova, S. Y., Khripach, L. V., and Shapot, V. S.,** Activity of soluble and bound hexokinase in transplantable hepatomas of various degrees of malignancy, and the normal and regenerating mouse liver (English transl.), *Dokl. Akad. Nauk S.S.S.R.,* 226, 1210, 1976.
204. **Gobeev, V. N. and Berezov, T. T.,** Biochemical method of estimating the degree of malignancy of transplantable hepatomas, *Dokl. Akad. Nauk S.S.S.R.,* English transl., 227, 750, 1976.
205. **Gobeev, V. N.,** Biochemical characteristics determining the rate of tumor growth in the organism (English transl.), *Biokhimiya,* 41, 2150, 1976.
206. **Sydow, G.,** Beziehungen zwischen hexokinaseaktivität und glykolyse in rattenleber und hepatomen, *Z. Naturforsh.,* 21, 232, 1966.
207. **Weinhouse, S., Shatton, J. B., Criss, W. E., and Morris, H. P.,** Molecular forms of enzymes in cancer, *Biochimie,* 54, 685, 1972.
208. **Lowry, O. H., Passonneau, J. V., Hasselberger, F. X., and Schulz, D. W.,** Effect of ischemia on known substrates and cofactors of the glycolytic pathway in brain, *J. Biol. Chem.,* 239, 18, 1964.
209. **Sacktor, B., Wilson, J. E., and Tiekert, C. G.,** Regulation of glycolysis in brain, *in situ,* during convulsions, *J. Biol. Chem.,* 241, 5071, 1966.
210. **Wilson, J. E., Sacktor, B., and Tiekert, C. G.,** *In situ* regulation of glycolysis in tetanized cat skeletal muscle, *Arch. Biochem. Biophys.,* 120, 542, 1967.

211. **Mandel, P.**, Free nucleotides in animal tissues, in *Progress in Nucleic Acid Research and Molecular Biology*, Vol. 3, Davidson, J. N. and Cohn, W. E., Eds., Academic Press, New York, 1964, 299.
212. **Veech, R. L.**, Freeze-blowing of brain and the interpretation of the meaning of certain metabolite levels, in *Cerebral Metabolism and Neural Function*, Passonneau, J. V., Hawkins, R. A., Lust, W. D., and Welsh, F. A., Eds., Williams & Wilkins, Baltimore, 1980, chap. 5.
213. **O'Sullivan, W. J.**, Stability constants of metal complexes, in *Data for Biochemical Research*, 2nd ed., Dawson, R. M. C., Elliott, D. C., Elliott, W. H., and Jones, K. M., Eds., Clarendon Press, Oxford, 1969, 423.
214. **Veloso, D., Guynn, R. W., Oskarsson, M., and Veech, R. L.**, The concentrations of free and bound magnesium in rat tissues, *J. Biol. Chem.*, 248, 4811, 1973.
215. **Weil-Malherbe, H. and Bone, A. D.**, Studies on hexokinase. I. The hexokinase activity of rat-brain extracts, *Biochem. J.*, 49, 339, 1951.
216. **Kipnis, D. M. and Cori, C. F.**, Studies of tissue permeability. V. The penetration and phosphorylation of 2-deoxyglucose in the rat diaphragm, *J. Biol. Chem.*, 234, 171, 1959.
217. **Rose, I. A. and O'Connell, E. L.**, The role of glucose 6-phosphate in the regulation of glucose metabolism in human erythrocytes, *J. Biol. Chem.*, 239, 12, 1964.
218. **Regen, D. M., Davis, W. W., Morgan, H. E., and Park, C. R.**, The regulation of hexokinase and phosphofructokinase activity in heart muscle, *J. Biol. Chem.*, 239, 43, 1964.
219. **Sauer, L. A.**, Hexokinase inhibition after the addition of glucose to Ehrlich-Lettre ascites tumor cells under aerobiosis, *J. Biol. Chem.*, 243, 2429, 1968.
220. **Uyeda, K. and Racker, E.**, Regulatory mechanisms in carbohydrate metabolism. VIII. The regulatory function of phosphate in glycolysis, *J. Biol. Chem.*, 240, 4689, 1965.
221. **Rose, I. A., Warms, J. V. B., and O'Connell, E. L.**, Role of inorganic phosphate in stimulating the glucose utilization of human red blood cells, *Biochem. Biophys. Res. Commun.*, 15, 33, 1964.
222. **Koobs, D. H.**, Phosphate mediation of the Crabtree and Pasteur effects, *Science*, 178, 127, 1972.
223. **Chou, A. C. and Wilson, J. E.**, Rat brain hexokinase: glucose and glucose-6-phosphate binding sites and C-terminal amino acid of the purified enzyme, *Arch. Biochem. Biophys.*, 165, 628, 1974.
224. **Wilson, J. E.**, Ligand-induced conformations of rat brain hexokinase: effects of glucose 6-phosphate and inorganic phosphate, *Arch. Biochem. Biophys.*, 159, 543, 1973.
225. **Siesjö, B. K.**, *Brain Energy Metabolism*, John Wiley & Sons, New York, 1978.
226. **Cori, C. F.**, Regulation of enzyme activity in muscle during work, in *Enzymes: Units of Biological Structure and Function*, Gaebler, O. H., Ed., Academic Press, New York, 1956, 573.
227. **Mayer, S. E., Nammi, D. H., and Hickenbottom, J. P.**, Regulation of the phosphorylase activating pathway in intact cardiac and skeletal muscle, in *Advances in Enzyme Regulation*, Vol. 8, Weber, G., Ed., Pergamon Press, New York, 1970, 205.
228. **Newsholme, E. A. and Start, C.**, *Regulation in Metabolism*, John Wiley & Sons, New York, 1973, chap. 4.
229. **Piros, R. and Staneloni, R.**, *In vivo* regulation of rat muscle glycogen synthetase activity, *Biochemistry*, 8, 2153, 1969.
230. **Pedersen, P. L. and Sacks, J.**, Hexosephosphate formation and the regulation of glycolysis in muscle, *Arch. Biochem. Biophys.*, 112, 548, 1965.
231. **Passonneau, J. V., Lowry, O. H., Schulz, D. W., and Brown, J. G.**, Glucose 1,6-diphosphate formation by phosphoglucomutase in mammalian tissues, *J. Biol. Chem.*, 244, 902, 1969.
232. **Rose, I. A. and Warms, J. V. B.**, Lithium action on glucose uptake in brain: role of glucose-1,6-P_2 as a regulator of hexokinase, *Biochem. Biophys. Res. Commun.*, 92, 1030, 1980.
233. **Beitner, R., Haberman, S., and Nordenberg, J.**, The effect of epinephrine and dibutyryl cyclic AMP on glucose 1,6-bisphosphate levels and the activities of hexokinase, phosphofructokinase and phosphoglucomutase in the isolated rat diaphragm, *Mol. Cell. Endocrinol.*, 10, 135, 1978.
234. **Beitner, R., Cohen, T. J., Nordenberg, J., and Haberman, S.**, Glucose 1,6-bisphosphate and the mechanism of the Pasteur effect in diaphragm muscle. The regulation of hexokinase and phosphofructokinase activities, *Biochim. Biophys. Acta*, 586, 266, 1979.
235. **Atkinson, D. E.**, The energy charge of the adenylate pool as a regulatory parameter. Interaction with feedback modifiers, *Biochemistry*, 7, 4030, 1968.
236. **Purich, D. L. and Fromm, H. J.**, Studies on factors influencing enzyme responses to adenylate energy charge, *J. Biol. Chem.*, 247, 249, 1972.
237. **Purich, D. L. and Fromm, H. J.**, Additional factors influencing enzyme responses to the adenylate energy charge, *J. Biol. Chem.*, 248, 461, 1973.
238. **Fromm, H. J.**, Control by energy charge is an untenable theory, *Trends Biochem. Sci.*, 2, N198-N200, 1977.
239. **Atkinson, D. E.**, Adenylate energy charge is a key factor, *Trends Biochem. Sci.*, 2, N198-N200, 1977.

240. **Pradet, A. and Raymond, P.,** Adenylate energy charge: concept and controversy, *What's New in Plant Physiol.*, 13, 21, 1982.
241. **Southard, J. H. and Hultin, H. O.,**The effect of energy charge on the activity of solubilized and bound mitochondrial hexokinase from chicken skeletal muscle, *Arch. Biochem. Biophys.*, 153, 468, 1972.
242. **Wilson, J. E.,** Ambiquitous enzymes: variation in intracellular distribution as a regulatory mechanism, *Trends Biochem. Sci.*, 3, 124, 1978.
243. **Knull, H. R., Taylor, W. F., and Wells, W. W.,** Effects of energy metabolism on *in vivo* distribution of hexokinase in brain, *J. Biol. Chem.*, 248, 5414, 1973.
244. **Bachelard, H. S.,** Biochemistry of coma, in *Biochemistry and Neurological Diseases*, Davison, A. N., Ed., Blackwell, London, 1976, 228.
245. **Knull, H. R., Taylor, W. F., and Wells, W. W.,** Insulin effects on brain energy metabolism and the related hexokinase distribution, *J. Biol. Chem.*, 249, 6930, 1974.
246. **Dirks, B., Hanke, J., Krieglstein, J., Stock, R., and Wickop, G.,** Studies on the linkage of energy metabolism and neuronal activity in the isolated perfused rat brain, *J. Neurochem.*, 35, 311, 1980.
247. **Bielicki, L. and Krieglstein, J.,** Solubilization of brain mitochondrial hexokinase by thiopental, *Naunyn-Schmiedeberg's Arch. Pharmacol.*, 298, 61, 1977.
248. **Bielicki, L. and Krieglstein, J.,** The effect of anesthesia on brain mitochondrial hexokinase, *Naunyn-Schmiedeberg's Arch. Pharmacol.*, 298, 229, 1977.
249. **Bruns, H., Krieglstein, J., and Wever, K.,** Narkose und intrazelluläre verteilung der cerebralen hexokinase, *Anaesthesist*, 27, 557, 1978.
250. **Hanke, J., Höfeler, H., Krieglstein, J., and Wever, K.,** Influence of various lipophilic drugs on brain mitochondrial hexokinase, *Naunyn-Schmiedeberg's Arch. Pharmacol.*, 307, 171, 1979.
251. **Höfeler, H. and Krieglstein, J.,** Solubilization of hexokinase activity by an effect of thiopental on the mitochondrial membrane, *Pharmacology*, 24, 156, 1982.
252. **Kyriazi, H. T. and Basford, R. E.,** An examination of rat brain hexokinase partitioning using "freeze-blown" tissue and synaptosomes, *Fed. Proc. Fed. Am. Soc. Exp. Biol.*, 42, 2079, 1983.
253. **Krieglstein, J., Schachtschabel, D. O., Wever, K., and Wickop, G.,** Influence of thiopental on intracellular distribution of hexokinase activity in various tumor cells, *Arzneim.-Forsch./Drug Res.*, 31, 121, 1981.
254. **Gumaa, K. A. and McLean, P.,** Sequential control of hexokinase in ascites cells, *Biochem. Biophys. Res. Commun.*, 35, 824, 1969.
255. **Gumaa, K. A. and McLean, P.,** A possible interrelationship between binding of hexokinase and the site of ATP formation in Krebs ascites cells, *Biochem. Biophys. Res. Commun.*, 36, 771, 1969.
256. **Moller, F. and Wilson, J. E.,** The influence of specific phospholipids on the interaction of hexokinase with the outer mitochondrial membrane, *J. Neurochem.*, 41, 1109, 1983.
257. **Ureta, T.,** The role of isozymes in metabolism: a model of metabolic pathways as the basis for the biological role of isozymes, in *Current Topics in Cellular Regulation*, Vol. 13, Horecker, B. L. and Stadtman, E. R., Eds., Academic Press, New York, 1978, 233.
258. **Nemat-Gorgani, M. and Wilson, J. E.,** Ambiquitous behavior — a biological phenomenon of general significance?, in *Current Topics in Cellular Regulation*, Vol. 16, Horecker, B. L. and Stadtman, E. R., Eds., Academic Press, New York, 1980, 45.

Chapter 4

HEXOKINASE IN HEALTH AND DISEASE

G. Rijksen and G. E. J. Staal

TABLE OF CONTENTS

I. INTRODUCTION

Hexokinase (ADP: D-hexose 6-phosphotransferase, EC 2.7.1.1) catalyzes the conversion of glucose to glucose-6-phosphate with $MgATP^{2-}$ as the phosphate donor, a step required for several glucose utilizing pathways. It is generally accepted that this reaction is one of the principle control points of glycolysis, together with the reaction catalyzed by phosphofructokinase (EC 2.7.1.11). Both reactions are attended with large changes in free energy, i.e., they are far displaced from equilibrium, and both enzymes are regulated by factors other than the substrate concentration. Therefore, they can be considered to be regulatory enzymes.[1] The involvement of hexokinase in the regulation of glucose utilization in a variety of diseases emphasizes its role as a regulatory enzyme of carbohydrate metabolism.

The in vivo activity of hexokinase is subject to regulation by environmental and hormonal stimuli resulting in increased or decreased (neo)-synthesis or destruction of particular isoenzymes. Furthermore, several intracellular factors are responsible for regulation of hexokinase activity, including its substrates $MgATP^{2-}$ and glucose, its products glucose-6-phosphate and MgADP, the important regulator of carbohydrate metabolism glucose-1,6-diphosphate, inorganic phosphate, and magnesium. Moreover, hexokinase exhibits a rather unique phenomenon in that it is regulated by its intracellular localization being a function of the metabolic status of the cell.[2]

In this review these regulatory mechanisms will be discussed in detail, with special reference to pathological conditions in which hexokinase plays a role.

II. ISOZYMIC SYSTEM OF MAMMALIAN HEXOKINASE

The isozymic pattern of mammalian hexokinase has been well documented since 1965 when Katzen and Schimke described the electrophoretic and kinetic properties of four isoenzymes of hexokinase from rat liver, designated as type I, II, III, and IV, respectively.[3] They can be distinguished by different substrate affinities and stabilities, apart from electrophoretic and chromatographic differences. Isoenzymes I, II, and III have a low K_m for glucose, rather broad sugar-substrate specificities and are inhibited by phosphorylated hexoses. Type III hexokinase, the isoenzyme with the highest affinity for glucose, shows substrate inhibition at higher concentrations of glucose. The main differences between the three low-K_m hexokinases concern their K_m for glucose and their in vivo and in vitro stabilities; their affinities for the nucleotide substrate do not differ mutually. Hexokinase type IV, the liver specific glucokinase, has a low affinity for glucose (K_m approximately 10 m*M*) and its sugar specificity is restricted to glucose; the enzyme is not inhibited by glucose-6-P. It is the only isoenzyme of hexokinase in liver which is regulated mainly by the concentration of glucose. *In situ* the low-K_m hexokinases of liver are fully saturated with glucose in all conditions and their activities will be regulated by several other factors as will be discussed later. The substrate inhibition observed for hexokinase III offers potentially a regulatory role for glucose as well. However, its physiological relevance remains unclear. Type III hexokinase is one of the main isoenzymes in liver, the major isoenzyme in granulocytes,[4] and is a minor component in several other tissues. It remains an intriguing question, why especially granulocytes are essentially exclusively equipped with a rather exceptional isoenzyme of hexokinase.

Type I is the most predominant form of hexokinase in all mammalian tissues. The presence of type II is characteristic for insulin-sensitive tissues, like muscle and adipose tissue, and its level is dependent on nutritional, hormonal, and developmental conditions (see Section V). As mentioned before, glucokinase is only found in liver and presumably functions during periods of high glucose supply and glycogen synthesis. Its level is under hormonal control as well. Detailed description of molecular and kinetic properties of the particular isoenzymes is outside the scope of this chapter and is subject of numerous reports in literature (see for reviews References 2, 5, and 6).

Each specific isoenzyme of hexokinase appears to be heterogeneous in nature. Two enzymatically active forms of the hexokinase type I could be isolated from pig heart.[7] It was documented that these forms originate from a reversible equilibrium between a dimer and a monomer form of the enzyme,[7,8] although in a later report the idea of a modification of one form into the other (e.g., proteolytic) was favored, each form being subject to dimerization.[9] The same monomer-dimer equilibrium has been described for bovine brain hexokinase.[10]

Hexokinase I from human erythrocytes consists of at least three subtypes,[11-13] which were isolated and characterized separately.[12,13] The three distinct forms do not differ in their affinity constants for the substrates glucose and $MgATP^{2-}$. In addition, the inhibition by glucose-1,6-diphosphate does not differ significantly. However, there appears to be significant differences with respect to the regulation of these inhibition by inorganic phosphate.[12] Aging of the red cell is accompanied by a relative increase of the proportion of the subtype which is less susceptible to regulation by inorganic phosphate.[12] The nature of the post-translational modification processes, which presumably underly these shifts, remains unclear.

Heterogeneity is a phenomenon common to the other isoenzymes of hexokinase as well. A distinct type II, called type II_f, was found in human fetal tissues[14] and in hepatomas.[15] A duplicate electrophoretic band of glucokinase can be found, especially when its activity is high.[16] In some instances duplicate bands of type III hexokinase can be found too, presumably caused either by enzyme polymorphism, or by in vitro modifications.[17]

The identification and nomenclature of multimolecular forms of hexokinase isoenzymes are mainly based on their electrophoretic mobilities. Therefore, they depend on the resolution of the electrophoretic system, which is used. As the mobilities of hexokinase I and II do not differ very much, some mixing up between subtypes of both isoenzymes in the literature may be present. Elucidation of the exact genetic origin requires in many instances isolation of the particular form and consecutive kinetic and immunological characterization.

III. REGULATION BY INTRACELLULAR FACTORS

A. Metabolites

Glucose-6-P has been recognized for a long time as a potent inhibitor of hexokinase in physiological important concentrations (see for reviews References 2, 5, and 6). Its inhibition offers a way of coordinate control of glycolysis by hexokinase and the other regulator enzyme of glycolysis, phosphofructokinase; regulation of phosphofructokinase will be reflected in variations of the level of glucose-6-P, which in turn affects the hexokinase activity. The interaction of hexokinase and glucose-6-P has been a matter of debate. The presence of a separate allosteric site for glucose-6-P was suggested, but recently it was established unequivocally that glucose-6-P binds to the active site of the enzyme[18] and thus is a competitive inhibitor with respect to $MgATP^{2-}$.[2,5,6]

More recently glucose-1,6-P_2 has drawn attention as an important regulator of carbohydrate metabolism.[19] Its importance is underlined by the following observations: (1) it exerts its influence on carbohydrate metabolism in physiological relevant concentrations, (2) its synthesis is under rigid control by several regulatory metabolites[20] and probably of hormonal stimuli as well, and (3) the levels of glucose-1,6-P_2 in muscle change under various physiological, hormonal, and pathological conditions, with concomitant changes in the activities of the enzymes which are affected by this regulator.[19] Glucose-1,6-P_2 is a potent inhibitor of hexokinase too[21-24] and acts competitively with respect to the substrate $MgATP^{2-}$.[24] At low concentrations of Mg^{2+} it is a much better inhibitor of hexokinase than glucose-6-P.[22] It may be interesting to note that hexokinase type II, which is a readily adaptive enzyme to several physiological conditions, concomitant with increases of the glucose-1,6-P_2 concentration (Section V), has a much greater affinity for the inhibitor than hexokinase I.[19]

The inhibition of hexokinase by glucose-1,6-P_2 may direct the cell to glycogenolytic activity, for, increasing levels of this metabolite cause a diminished rate of glucose phosphorylation, but stimulates glycolysis by stimulating phosphofructokinase activity.[19] Several other metabolites are able to inhibit the hexokinase reaction, e.g., the other product of the reaction $MgADP^-$, MgAMP, fructose-6-P, 2,3-diphosphoglycerate, and 3-phosphoglycerate.[25] However, comparison of the inhibition constants with intracellular concentrations renders a physiological significance very unlikely with the possible exception of 2,3-diphosphoglycerate in erythrocytes.[25-28]

B. Inorganic Phosphate

Inorganic phosphate (P_i) has no direct influence on the hexokinase reaction, but it is known to be involved in the regulation of hexokinase activity by a relieving effect on the inhibition by glucose-6-P[25,29-32] and glucose-1,6-P_2.[24] Stimulation of hexokinase activity by P_i is paralleled with P_i stimulation of phosphofructokinase resulting in an increase of the glycolytic flux.[33] Kinetic studies of the P_i action on the inhibition by glucose-6-P and glucose-1,6-P_2 offers a model which implies that hexokinase may exist in equilibrium either as a free or a phosphate-associated enzyme, the kinetic parameters of the two enzymes being similar with exception of the binding constant for P_i.[24,25,30-32] Alternatively, this model can be modified by assuming the presence of two fundamental different enzymes, presumably generated by posttranslational modification, differing in their ability to bind P_i.[24,25,34] The latter was shown to be the case in erythrocytes, from which several subtypes of hexokinase type I could be isolated with different affinities for P_i.[12]

C. Magnesium Ions

The influence of Mg^{2+} is merely attributed to its complexing properties. The true substrate in the hexokinase reaction is $MgATP^{2-}$, while free ATP^{4-} is inhibitory.[35-37] However, at intracellular concentrations of free Mg^{2+}, estimated to be approximately 0.7 m*M* in erythrocytes[38] and approximately 1.5 m*M* in brain,[35] the concentration of ATP^{4-} is minimal, so that its inhibitory action will be negligible. Mg^{2+} might be involved in the regulation of hexokinase too by complexing with glucose-1,6-P_2. Unlike the inhibition by glucose-6-P, that of the diphosphate was quite sensitive to the presence of physiological concentrations of free Mg^{2+}; the Mg-Glucose-6-P complex was not inhibitory. The inhibition constant for this inhibitor was increased three times by increasing free Mg^{2+} from 0.5 to 5.0 m*M*.[39]

D. pH

The regulatory role of pH in glycolysis was only attributed to its effect on the enzyme phosphofructokinase[33] and not to an effect on hexokinase. Indeed, hexokinase is not directly influenced by small pH changes, because it has a very broad pH optimum (7.0 to 8.4) and because the kinetic properties do not vary with pH.[40] However, there is an indirect effect on the activity of hexokinase by influencing the inhibition by glucose-1,6-P_2. Increase of pH increases exponentially the K_i for glucose-1,6-P_2.[22,24] The observed pH effects are ascribed to the phosphate group at the C-1 atom which is bound as a monoanion, but is excluded as a dianion.[22] The physiological importance, however, remains obscure, but may emerge in tissues in which relative great variations in pH are known to occur, e.g., in muscle.

E. General Considerations

The intracellular factors that might regulate hexokinase activity are summarized in Figure 1. It seems obvious that the real in vivo situation with respect to hexokinase activity is not

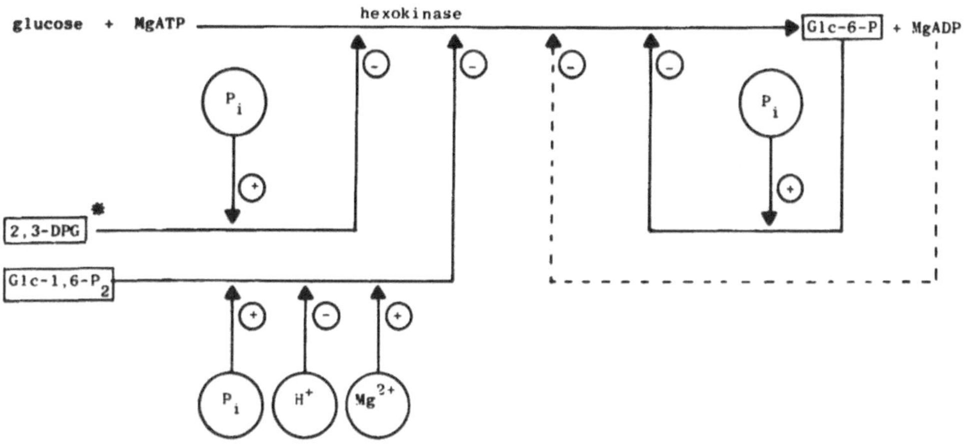

FIGURE 1. Factors involved in the regulation of red cell hexokinase activity, either directly or indirectly. ⊕ or ⊖ denote direct or indirect stimulatory or inhibitory effect, respectively, on the phosphorylation of glucose. *, in erythrocytes only.

simply the sum of these factors and the amount of enzyme protein present; their interplay and their "in vivo" state must be considered.

Hexokinase can occur either in soluble form or bound to mitochondria and its activity is regulated by the soluble/particulate distribution (Section IV). Also regulatory metabolites may be particle bound. Therefore, they may be either not available for hexokinase regulation or be of particular interest by localization in the ultimate of the particulate enzyme. In erythrocytes 2,3-diphosphoglycerate, ATP, glucose-6-P, and P_i can be bound to hemoglobin dependent on the oxygenation state.[38,41-45] In blood platelets, in which essential all hexokinase is bound, a large proportion of glucose-6-P is particulate as well.[46] The role of compartmentation in the regulation of hexokinase will be discussed in detail in Section IV.

IV. ROLE OF COMPARTMENTATION

Hexokinase from several tissues can be bound reversibly to the mitochondrial membrane. It can be solubilized from the membrane by physiologically relevant concentrations of glucose-6-P, glucose-1,6-P_2, and ATP; the binding is promoted by Mg^{2+} and ADP.[2] Indeed the proportion of hexokinase associated with the mitochondria in brain is dependent on cerebral energy status and increases during periods of glycolytic stress.[2,47,48] To describe such behavior the term "ambiquitous enzyme" has been suggested, which is defined more generally as one "whose distribution between cytoplasmic and particulate (membrane-bound) forms may vary with metabolic status of the cell as reflected in the levels of certain metabolites capable of influencing that distribution."[2,49] When bound to mitochondria the enzyme is less susceptible to inhibition by glucose-6-P and glucose-1,6-P_2[50,51] and may have a greater affinity for its substrate $MgATP^{2-}$ than the soluble enzyme (the latter is still a matter of dispute).[2] So, binding to the mitochondria would facilitate anaerobic glycolysis. Indeed, Kurakowa et al.[52] showed that the bound hexokinase is almost twice as effective in terms of glucose-6-P production as the unbound enzyme. Moreover, it has been suggested that bound hexokinase can directly benefit from the ATP produced by the mitochondrial respiration for the phosphorylation of glucose.[53] Accordingly, the predominance of mitochondrial bound hexokinase is found in tissues which are critically dependent on glucose metabolism like brain,[2] blood platelets,[4,46] lymphocytes,[4] highly glycolytic regions of the kidney,[54] and neoplastic tissues (see Section VI). In a series of hepatomas it was shown that the amount of mitochondrial hexokinase correlated with growth rate and glycolytic activity.[55] In a comparative study on several rat tissues increased amounts of hexokinase appeared to be associated with the mitochondria in tissues, which were less dependent on glycogen metabolism.[56] Inversely, cells exhibiting predominantly cytosolic hexokinase might be particularly suitable for the use of endogeneous glycogen as a carbon source.[56,57] The reason might be the greater sensitivity for the inhibition by glucose-1,6-P_2, which concomitantly can increase the glycolytic flux by stimulating phosphofructokinase activity.

There are two main hypotheses regarding the presence of cytosolic and mitochondrial forms of hexokinase.[2] First, there is a simple binding equilibrium, dependent on the metabolic status of the cell. Second, the two forms are fundamentally different. Several identical properties of the soluble and solubilized enzyme,[51,58-60] such as molecular weight, chromatographic behavior, electrophoretic mobility, and the main kinetic properties clearly suggest that both forms are basically identical proteins. However, some authors report rather subtle differences in kinetic behavior,[57,61] which indicate that distinct cytoplasmic and mitochondrial forms may exist. Furthermore, developmental changes in the soluble/particulate distribution in cerebellum and cerebrum as well as regional differences in adult brain have been used as arguments in favor of the presence of fundamental differences between soluble and bound hexokinase.[2] It has been reported that the predominant form of hexokinase in astrocytes and glial cells is cytosolic (note that astrocytes use considerable amounts of endogeneous glycogen, which is in line with the aforementioned hypothesis),[57] while virtually all hexokinase in neuronal cells is particulate.[57,62]

However, recently it was demonstrated that soluble and solublized hexokinase from rat brain can bind equally well to depleted brain mitochondria or to rat liver mitochondria, which observation strongly suggests that both enzymes are one and the same enzyme in simple equilibrium with each other.[31] The different localization of hexokinase in glial cells and neuronal cells would thus be either the reflection of a different metabolic state in these cells or an intrinsic property of the respective mitochondria. As limited proteolysis in vitro has been shown to destroy the binding properties of both soluble and solubilized brain enzyme,[51,63] the soluble/particulate distribution in other tissues may well be influenced by intracellular proteolysis or other posttranslational modification processes. In liver essentially all hexokinase is cytosolic[56] and it cannot be bound under in vitro conditions either.[56,64] Yet, liver mitochondria are capable of binding high amounts of hexokinase from other sources,[65] which argues for a posttranslational modification of the liver enzyme destroying its ability to bind to mitochondria.

A comprehensive review on the soluble and particulate hexokinase type I and the factors regulating their compartmentation in vivo and in vitro is given in Reference 2. Unlike type III hexokinase,[4,66] type II can also bind to mitochondria,[67,68] but much less information is available comparing the soluble and particulate form of the enzyme. Studies on hexokinase II from ascites tumor cells indicate differences between the bound and soluble form with respect to the inhibition by phosphorylated hexoses as well[67] and show that the enzyme, which is rather labile in soluble form, is greatly stabilized by binding.[68]

In the binding of hexokinase to mitochondria a more or less specific hexokinase-binding protein in the outer membrane is involved, which has been purified and characterized as a protein with a subunit molecular weight of 31 kD.[49,65] Recently,[69] it has been identified as the protein porin, which is known to be a channeling protein in the outer mitochondrial membrane of rat liver. ADP-ATP exchange with the cytosol across the outer membrane is only possible through the pore formed by the mitochondrial porin. This identification of the hexokinase-binding protein links mitochondrial bound hexokinase immediately to the ADP-ATP translocator in the membrane, which strongly supports the hypothesis that the enzyme might use mitochondrial-borne ATP for the phosphorylation of glucose.

V. DIETARY AND HORMONAL FACTORS IN THE REGULATION OF HEXOKINASE ACTIVITY

Glucose phosphorylation in vivo is regulated by a number of environmental factors provided by dietary and hormonal stimuli. Hexokinase type II and IV (glucokinase), and probably type III, are adaptive to several physiological states, while type I hexokinase does not seem to be adaptive.

In cultured hepatocytes glucokinase activity could be increased by the addition of insulin or dexamethasone, apparently as the result of new enzyme biosynthesis. Addition of glucagon reduced the level of glucokinase.[70] Starvation of rats induced a decay of glucokinase activity in liver, the activity being restored after refeeding.[71,72] Diminished insulin levels were shown to be responsible for the reduction in glucokinase activity during fasting, while, on the other hand, glucagon, catecholamines, and cAMP could prevent it.[72] In addition, glucocorticoids, thyroid hormone, and cGMP are supposed to play a role in the insulin-mediated regulation of glucokinase activity.[70,73] The data discussed above were obtained from rat livers after fasting and refeeding or from alloxan-diabetic rats. Diabetes in humans is accompanied with a much lesser decrease of glucokinase activity which suggests a different regulation mechanism for rats and humans.[74]

Insulin causes a similar increase of hexokinase type II activity. In conditions of low insulin secretion, e.g., during fasting or in diabetes, hexokinase II is low in insulin-sensitive tissues,[75-78] returning to normal when the insulin level is normalized. The absolute level of hexokinase II depends on the carbohydrate content of the diet. Furthermore, a close correlation exists between glucose utilization in the presence of insulin and the activity of hexokinase II in adipocytes, which suggests that HK II mediates the hormonal action of insulin on glucose utilization.[78] The results on human tissues roughly equal the data obtained with rat alloxan-diabetic rats.[79]

The exact mechanism of the adaptation of hexokinase activity to hormonal stimuli still remains unknown and the involvement of direct effects of variations in carbohydrate metabolites should be considered also. In adipocytes the activity of hexokinase type II can be increased in vitro by the addition of glucose or other metabolizable hexoses in the absence of insulin.[78] Addition of insulin potentiates this effect. In contrast, the amount of hexokinase type II in cultured B-lymphocytes inversely correlates with the glucose concentration in the medium.[80] Refreshment of culture medium switches off the neosynthesis of the enzyme, while exhaustion of the medium causes a concomitant restoration of enzyme activity.

In addition to dietary and hormonal effects on the concentrations of enzyme protein, regulatory factors of hexokinase activity may be under hormonal control too. The mitochondrial binding of hexokinase, which is supposed to be an important factor in regulating hexokinase activity, is promoted by insulin.[77,81] Alternatively, diabetes produces a release of particulate hexokinase activity,[75,82] rendering less active enzyme (see Section IV). In this light the "hexokinase acceptor theory" should be mentioned, which hypothesizes that insulin could act as a bridge between hexokinase and the mitochondrial membrane in order to produce an increase in the respiratory control.[83] In other words there might exist a close relationship between insulin and the hexokinase binding protein in the mitochondrial membrane. However, definite proof supporting this hypothesis is lacking.

As pointed out above, glucose-1,6-P_2 is an important regulator of hexokinase activity too. The level of glucose-1,6-P_2 in muscle fluctuates under different physiological conditions, leading to concomitant changes in the in vivo activities of hexokinase type I and II.[19] Several conditions which affect the level of glucose-1,6-P_2 are summarized by Beitner in Reference 19 and Chapter 1 in this book. It should be noticed that during fasting or in diabetes the level of glucose-1,6-diphosphate is decreased, which will result in an increased activity of hexokinase type I and II. However, this effect will be counteracted at a longer term by the decrease of enzyme protein concentration — at least of type II hexokinase — in hypoinsulinemic states. Similarly, the direct effects of hormonal action on enzyme neosynthesis and its indirect effect on enzyme regulation via the level of glucose-1,6-P_2 during insulin administration or refeeding (after fasting) act in an opposite direction.

VI. HEXOKINASE DEFICIENCY

The erythrocyte is a metabolically restricted cell, which depends heavily on glycolysis for the generation of its energy demands. It is generally accepted that hexokinase and phosphofructokinase are the principle control points in red cell glycolysis. The relative importance of these two enzymes could be assessed by mathematical modeling of erythrocyte glycolysis.[41] It appeared that the "control strength" of hexokinase — defined as the ratio of the change of the glycolytic flux induced by an effector over the change in particular enzyme activity — was about twice the control strength of phosphofructokinase at physiological pH.

The importance of the hexokinase reaction for erythrocyte metabolism is further illustrated by the manifestation of severe nonspherocytic hemolytic anemia (NSHA) in patients with a partial deficiency of this enzyme. The first association of hexokinase deficiency and NSHA was described by Valentine et al.[85] The kinetic and electrophoretic properties of the enzyme showed no abnormalities. Since this report, about ten other families with hexokinase deficiency and NSHA were reported as recently reviewed by Paglia et al.[86] The mutant enzymes showed either no abnormal properties at all,[87,88] or decreased substrate affinities[89-92] or heat stabilities,[89,93,94] while in some patients altered regulatory properties were detected.[34,95] In the latter cases the enzyme was less inhibited by phosphorylated hexoses and there were slight alterations in the regulation of these inhibitions by inorganic phosphate in one of these patients.[34] Electrophoretically the deficiency could be attributed to one particular subtype of hexokinase type I. Most other variants showed no abnormal electrophoretic mobilities, except in the case described by Necheles et al.,[90] in which the deficiency of hexokinase was characterized kinetically and electrophoretically by the absence of type III hexokinase, while hexokinase I was present in normal amounts. The type III isoenzyme is present in low concentrations in normal erythrocytes, hexokinase I being the most predominant isoenzyme. In all other patients reported, the deficiency affected the type I isoenzyme.

The diagnosis of hexokinase deficiency in NSHA is usually troubled by the presence of very young cell populations with reticulocyte counts up to 70 to 80%. Hexokinase is an age-dependent enzyme and its activity in mature erythrocytes is estimated to be less than 10% of that in mature erythrocytes.[86,88,94] There must be a rapid decrease of activity during maturation or a sudden decay during the transition of the reticulocyte to a mature erythrocyte by posttranslational modification reactions. This decrease in enzyme activity is accompanied by the appearance of several subtypes of hexokinase I in the electrophoretic pattern,[11-13] concomitant with subtle changes in regulatory properties.[12] In patients with NSHA due to a hexokinase defect the activity of the enzyme is usually only recognized to be deficient when the degree of reticulocytosis is taken into account. Due to these high reticulocyte counts the impact of hexokinase deficiency on cell metabolism should be considered in relation to the metabolic demands of the reticulocyte, which is metabolically much more complex than the mature erythrocyte.[96] Furthermore, in reticulocytes hexokinase is probably bound to mitochondria,[97] which are destroyed during the transition to mature erythrocytes. The metabolic events during this process and the importance of hexokinase in the generation of energy in these stages are unknown. It is conceivable that the cell is particularly vulnerable to hemolysis during certain stages of this process in hexokinase deficiency.

The study of hexokinase deficiency has been restricted mostly to the red cell. However, as the deficiency concerns the most abundant isoenzyme type I, it probably affects other cell types too. Yet, hexokinase deficiency manifests itself primarily in NSHA, although in some patients a diversity of abnormalities, including multiple malformations, panmyelopathy, and latent diabetes mellitus have been described.[98,99] The enzyme defect in these patients, however, appeared to be acquired, rather than truly genetic.

Blood platelets[100] and lymphocytes[101,102] both depend more or less heavily on energy supplied by glycolysis and both contain type I hexokinase as the major isozyme.[4] Hexokinase I is therefore a likely candidate to play an essential role in the energy metabolism in these cells too. In a patient with NSHA and red cell hexokinase deficiency recently described by Rijksen et al.[95] a diminished hexokinase activity was detected also in blood platelets (20 to 35% of normal activity). Some properties of the platelets, e.g., Ca-secretion, were disturbed upon metabolic stress in vitro.[103] Although the total hexokinase activity in lymphocytes was normal, the amount of hexokinase type I was decreased to about 50% of normal.[95] However, the deficiency was compensated by the appearance of type III hexokinase. Remarkably, compartmentation studies showed that this type III enzyme was localized in the cytosol, while essentially all hexokinase activity in normal lymphocytes is particulate.[4] No abnormal lymphocyte functions could be detected.[95] In some pathological conditions other than merely NSHA the presence of a hemolytic process might be caused by a secondary hexokinase deficiency. Human red cell hexokinase can be inhibited profoundly by elevated concentrations of copper.[104] Serum copper levels and red blood cell free copper are significantly increased in some patients with malignant lymphoma, and it has been suggested that the shortened red cell life span often observed in these patients results from a copper-induced inhibition of red cell hexokinase.[105] The same mechanism may be responsible for the hemolysis observed in patients with Wilson's disease, as high copper levels are known to occur in the tissues of these patients.[106]

VII. HEXOKINASE IN NEOPLASIA

In general neoplastic tissues have been characterized by the presence of high glycolytic capacity as was recognized as early as 1930 by Warburg.[107] An increase in glycolytic flux can only be accomplished by concomitant changes in the in vivo activities of the regulatory enzymes of glycolysis. This may involve increased synthesis or decreased degradation of enzyme protein, change of the isoenzyme composition, thereby altering the regulatory properties, or changing levels of intracellular regulatory factors. Most of the studies on key enzymes of glycolysis in neoplasia were performed on rat hepatomas, because they provide a spectrum of well-defined tumors ranging from slowly growing, well differentiated to rapidly proliferating undifferentiated tumors. In general, a very good correlation was observed between hexokinase activity and growth rate or differentiation rate of the tumors,[15,16,53,55,108-110] while an inverse correlation was found with the glucokinase activity.[16,108] Highly differentiated hepatomas closely resemble normal liver with respect to hexokinase activity and isoenzyme composition.[16] On the contrary, poorly differentiated hepatomas are similar to fetal liver in these respects.[16,108]

The increase in hexokinase in less differentiated hepatomas is accompanied by a change in isoenzyme composition. In normal liver all four isoenzymes are present, type I and IV being most predominant. Dedifferentiation accomplishes a shift in favor of either type II[15,16,108] or type III hexokinase.[16,109] The reports are rather conflicting in this respect. According to some authors in relatively well-differentiated Morris hepatomas relatively predominant type II hexokinase is found in addition to type I and III, resembling the pattern in regenerating liver and fetal liver.[15,108] In poorly differentiated Yoshida ascites hepatomas only type I and a markedly elevated type II are present, with no detectable type III and IV hexokinase.[108] In contrast, in a series of transplantable tumors in rat, including hepatomas, Farron showed that all tumors contained significant amounts of hexokinase III, whether or not their respective tissue of origin expressed this isoenzyme.[109] Hexokinase III was absent from fetal tissues and thus served to distinguish neoplastic tissues from normal undifferentiated tissues. In another study both type II and type III hexokinase were shown to be elevated in poorly differentiated hepatomas.[16] In the appreciation of these rather conflicting results it is worth noting that in these studies only cytosolic hexokinase activity was measured. In hepatomas a considerable part of hexokinase is particulate,[53,55,110-113] as will be discussed below. Therefore, the proportion of the actual increases of the particular isoenzyme activities may be masked by differences in the degree of mitochondrial binding dependent on the experimental conditions.

Increased hexokinase activity and the appearance of hexokinase type II is a finding common to a number of other tumors too,[114-117] but yet it cannot be considered to be a general phenomenon. In contrast, hexokinase activity in some tumors of the brain, e.g., gliomas, astrocytomas,[118] neuroblastomas,[119] and retinoblastomas[120] was markedly decreased with respect to normal brain, while in most of these tumors only type I hexokinase could be detected.[118] This finding may be related to the highly glycolytic nature of adult brain compared to other tissues.[2]

As hexokinase type II is a very adaptive enzyme, as discussed in Section V, the appearance of this particular isoenzyme in tumors may well be provoked by altered dietary and hormonal conditions or other environmental factors (e.g., vascularization), rather than being primarily the expession of malignant transformation or increased proliferation. Support for the former proposition is governed by the observation that in rats bearing nonhepatic tumors, hexokinase activity in liver increases significantly and the isozymic composition shifts towards the pattern found in immature liver.[121,122] These biochemical responses to distant neoplasia occur within 24 hr after the transplantation of the tumor and are therefore a very early sign of neoplasia. The same alterations in hepatic enzyme pattern were seen in humans with distant neoplasia.[122] These observations have led to the hypothesis of a circulating humoral factor inducing a dedifferentiation in the host metabolism.[121]

The in vivo activity of hexokinase can be regulated by the intracellular compartmentation, as discussed in Section IV. In hepatomas a considerable amount of hexokinase activity is associated with the mitochondria. This important phenomenon was originally reported by several workers[111-113] and studied in detail by Bustamante and Pedersen.[53,55,110] In a series of hepatomas the amount of bound hexokinase appeared to correlate with the growth rate and the glycolytic activity of the tumors.[55] The tumor mitochondrial hexokinase was directly coupled to oxidative phosphorylation, because addition of glucose to respiring hepatoma mitochondria (after a burst of ATP synthesis) resulted in stimulation of respiration,[53] whereas glucose had no effect on the respiration of mitochondria from control and regenerating liver, which bear no hexokinase on their membranes. Addition of tumor mitochondria to normal rat liver cytosol produced a stimulatory effect on glycolysis, while tumor mitochondria devoid of hexokinase activity did not.[55] From these studies it has been concluded that mitochondrial bound hexokinase might be responsible, at least in part, for the high aerobic glycolytic flux in hepatomas. This conclusion was further supported by the finding that substitution of glucose in the culture medium, by galactose, which can enter the glycolytic chain by the Leloir-pathway bypassing the hexokinase step, produced a marked decrease in lactic acid production. Remarkably, the cells grew equally well in both conditions.[53] However, it was not established unequivocally whether mitochondrial bound hexokinase or increased hexokinase activity per se was stimulatory for tumor glycolysis. Moreover, the possibility that the high glycolytic rate in tumor cells was provoked by an ineffective Na-K-ATPase, as stated by others,[123] is bypassed by the use of rather high concentrations of inorganic phosphate in some crucial experiments.

In addition, the correlation between high levels of mitochondrial bound hexokinase and tumor growth rate and glycolytic activity cannot be considered to be a general phenomenon in cancer tissues. Comparative studies on rapidly growing undifferentiated medullary thyroid carcinomas in the rat vs. slowly growing well-differentiated tumors of the same origin showed no differences in either hexokinase specific activity or in the rate of compartmentation in spite of a significant increase in lactic acid production in rats bearing the undifferentiated tumor with respect to the well-differentiated tumor-bearing animals.[50] Furthermore, in human astrocytomas and gliomas the rate of compartmentation of hexokinase was decreased compared to normal brain.[118]

There is hardly any evidence available in the literature for the presence of kinetic and/or regulatory differences which can be contributed to tumor specific alterations and which cannot be found in any normal tissue. One possible exception is governed by the studies of Johnson et al.,[124] who showed that glycolysis in Ehrlich ascites tumor cells could be inhibited markedly by the addition of a chloromethyl-ketone derivative of lactic acid, 2-isobutyl 3-oxo-4-chloro-2-butyl-carbonate. The inhibition could be localized at the hexokinase and phosphofructokinase steps and acted probably by interfering with essential SH groups. Hexokinase from mouse brain required 50 times higher concentrations for inhibition than the enzyme from mouse Ehrlich ascites tumors cells. Glycolysis and hexokinase activities in extracts of Rous sarcoma virus transformed cells were considerably more sensitive to the inhibitor than the activities from normal chick embryo fibroblasts.

As not only tumor glycolysis, but also cell growth was inhibited by the lactic acid derivative, this finding is of potential therapeutic interest. The origin of the observed differences in behavior between tumor hexokinase and normal hexokinase may be found in a different isozymic composition, although iso-electric focusing revealed no major differences between mouse Ehrlich ascites cells and mouse brain, or from differences in intracellular compartmentation. However, one of the most attractive possibilities is a specific alteration caused by a transformation-linked posttranslational modification, as it would render a tumor specific goal for therapeutic approaches.

VIII. ROLE OF HEXOKINASE IN OTHER PATHOLOGICAL CONDITIONS

A. Diabetes

The involvement of hexokinase in diabetic states has been discussed in an earlier section on the hormonal action of insulin. The available evidence suggests that the insulin-dependent control of glucose metabolism is mediated by the level of type II hexokinase, the isoenzyme which predominates in insulin-sensitive tissues.

Type I hexokinase is not regulated by hormonal control and is most predominant in insulin-insensitive tissues. Yet, some disturbances in glucose phosphorylation have been detected in insulin-insensitive tissues of diabetic subjects. In erythrocytes of diabetic patients the level of glucose-6-P was increased and was suggested to be correlated with the level of glycosylated hemoglobin.[125] As red cells are fully permeable to glucose, hexokinase will be fully saturated and no alterations in activity are to be expected when the glucose concentration is elevated. Yet, besides glucose-6-P no other glycolytic intermediates were found to be aberrant, which argues for a modified glucose phosphorylation in these cells.[126] The proximate cause of this alteration is not known, but may well involve a decreased level of glucose-1,6-P_2, which has been shown to be low in other diabetic tissues.[19] The metabolism of glucose-1,6-P_2 is probably under the hormonal control of insulin[19] and variations in its level may thus contribute to alterations in glucose phosphorylation in diabetes.

B. Muscular Dystrophy

In the muscle of patients with X-linked dystrophies of Duchenne and Becker the isozymic pattern of hexokinase is changed, while the total hexokinase activity does not differ significantly from normal.[127,128] In the electrophoretic pattern the normal hexokinase type II band is absent and is replaced by bands having electrophoretic mobilities different from those of normal adult or fetal muscle. In patients with a myopathy of other etiology no such aberrations could be detected.[129] It was argued that the change in isoenzyme composition might underly the observation that in muscle of patients with dystrophy of Duchenne more glucose than normal is converted to fructose instead of to glucose-6-P. The latter seems to be in contradiction with the observation that in muscle of dystrophic mice the level of the hexokinase inhibitor glucose-1,6-P_2 is strikingly decreased with a concomitant increase of hexokinase type II activity.[19,130,131] These observations, however, may be brought in line by determination of the levels of glucose-1,6-P_2 in muscle of patients with dystrophy of Duchenne or Becker and by exploring the possibility that the alteration of the electrophoretic pattern in these patients is accompanied by a change in the regulation of type II hexokinase by glucose-1,6-P_2.

C. Wiskott-Aldrich Syndrome

Wiskott-Aldrich syndrome (WAS) is a rare sex-linked primary immunodeficiency, which is characterized by thrombocytopenia and eczema in addition to frequent infections. In two patients with WAS a partial deficiency of hexokinase (30% of normal activity) was detected in the lymphocytes, while the activity in erythrocytes was normal.[132] Investigations in another patient revealed a partial deficiency in platelet hexokinase, while in lymphocytes and erythrocytes the activity was normal.[133] In lymphocytes, blood platelets as well as erythrocytes the primary isoenzyme is type I hexokinase. Therefore, a partial deficiency in only one of these cell types can hardly be expected to result from a proximate defect at the hexokinase locus, but rather reflects a secondary change due to an altered intracellular metabolism in the particular cell type. In patients with genetically hexokinase type I deficiency, the prime manifestation is severe hemolytic anemia, while no disturbance in the immune system can be detected.[98]

D. Liver Diseases

In patients with hepatitis or liver cirrhosis hexokinase activity in the liver is increased with a concomitant decrease in glucokinase activity.[134,135] There is a marked increase in type III hexokinase in the acute phase of hepatitis and a gradual increase in type I hexokinase as the disease advances from chronic hepatitis to cirrhosis of the liver.[134] Histochemical investigations showed that the increased hexokinase activity is mainly localized in parenchymal cells which revealed evidence suggestive of the presence of cell injuries.[135]

Acute hepatic injury induced in rat liver by CCl_4 causes a similar increase of liver hexokinase and a similar isozymic shift.[136] It can be concluded that the predominance of type III hexokinase together with an increased total activity in liver is indicative for acute hepatic injury.

E. Anesthesia

During anesthesia the rate of glucose utilization in brain is diminished and it has been shown that this can be the result of an interference on the compartmentation of hexokinase by anesthetics.[137,138] The anesthetic thiopental in therapeutically relevant concentrations causes solubilization of cerebral hexokinase from the mitochondrial membrane and therefore inhibits phosphorylation of glucose by shifting hexokinase activity from its more active (mitochondrial) form to its less active (soluble) form. The action of thiopental is probably direct, as the solubilization of mitochondrial hexokinase by the drug can be achieved from isolated mitochondria.[139] This phenomenon should be kept in mind when compartmentation studies are performed on tissue samples obtained during surgery.

REFERENCES

1. **Newsholme, E. A. and Start, C.,** *Regulation in Metabolism*, John Wiley & Sons, London, 1973, chap. 1.
2. **Wilson, J. E.,** Brain hexokinase; the prototype ambiquitous enzyme, in *Current Topics in Cellular Regulation*, Vol. 16, Horecker, B. L. and Stadtman, E. R., Eds., Academic Press, New York, 1980, 1.
3. **Katzen, H. M. and Schimke, R. T.,** Multiple forms of hexokinase in the rat: tissue distribution, age dependency, and properties, *Proc. Natl. Acad. Sci. U.S.A.*, 54, 1218, 1965.
4. **Rijksen, G., Staal, G. E.J., Beks, P. J., Streefkerk, M., and Akkerman, J. W. N.,** Compartmentation of hexokinase in human blood cells. Characterization of soluble and particulate enzymes, *Biochim. Biophys. Acta*, 719, 431, 1982.
5. **Colowick, S.P.,** The hexokinases, in *The Enzymes*, Vol. IXB, Boyer, P. D., Ed., Academic Press, New York, 1973, 1.
6. **Purich, D. L., Fromm, H. J., and Rudolph, F. B.,** The hexokinases: kinetic, physical, and regulatory properties, in *Advances in Enzymology*, Vol. 39, Meister, A., Ed., John Wiley & Sons, New York, 1973, 249.
7. **Easterby, J. S. and O'Brien, M. J.,** Purification and properties of pig-heart hexokinase, *Eur. J. Biochem.*, 38, 201, 1973.
8. **Easterby, J. S.,** Heart hexokinase: quaternary structure changes accompanying the binding of regulatory molecules, *Eur. J. Biochem.*, 58, 231, 1975.
9. **Vowles, D. J. and Easterby, J. S.,** Comparison of type I hexokinases from pig heart and kinetic evaluation of the effects of inhibitors, *Biochim. Biophys. Acta*, 566, 283, 1979.
10. **Chakrabarti, U. and Kenkare, U. W.,** Dimerization of brain hexokinase induced by its regulator glucose-6-phosphate, *J. Biol. Chem.*, 249, 5984, 1974.
11. **Rijksen, G., Schoop, I., and Staal, G. E. J.,** Properties of human erythrocyte hexokinase related to cell age, *Clin. Chim. Acta*, 80, 193, 1977.
12. **Rijksen, G., Jansen, G., Kraaijenhagen, R. J., van der Vlist, M. J. M., Vlug, A. M. C., and Staal, G. E. J.,** Separation and characterization of hexokinase I subtypes from human erythrocytes, *Biochim. Biophys. Acta*, 659, 292, 1981.
13. **Stocchi, V., Magnani, M., Canestrari, F., Dacha, M., and Fornaini, G.,** Multiple forms of human red blood cell hexokinase. Preparation, characterization and age dependence, *J. Biol. Chem.*, 257, 2357, 1982.
14. **Cayanis, E. and Balinsky, D.,** Comparative kinetic properties of human hexokinases, *Int. J. Biochem .*, 6, 741, 1975.
15. **Hammond, K. D. and Balinsky, D.,** Isozyme studies of several enzymes of carbohydrate metabolism in human adult and fetal tissues, tumor tissues and cell cultures, *Cancer Res.*, 38, 1323, 1978.
16. **Shatton, J. B., Morris, H. P., and Weinhouse, S.,** Kinetic, electrophoretic and chromatographic studies on glucose-ATP phosphotransferases in rat hepatomas, *Cancer Res.*, 29, 1161, 1969.
17. **Povey, S., Corney, G., and Harris, H.,** Genetically determined polymorphism of a form of hexokinase, HK III, found in human leucocytes, *Ann. Hum. Genet. London*, 38, 407, 1975.
18. **Solheim, L. P. and Fromm, H. J.,** Kinetic evidence that the high-affinity glucose 6-phosphate site on hexokinase I is the active site, *Arch. Biochem. Biophys.*, 211, 92, 1981.
19. **Beitner, R.,** The role of glucose-1,6-bisphosphate in the regulation of carbohydrate metabolism in muscle, *Trends Biochem. Sci.*, 4, 228, 1979.
20. **Rose, I. A., Warms, J. V. B., and Wong, L.-J.,** Inhibitors of glucose-1,6-biphosphate synthase, *J. Biol. Chem.*, 252, 4262, 1977.
21. **Gerber, G., Preissler, H., Heinrich, R., and Rapoport, S. M.,** Hexokinase of human erythrocytes. Purification, kinetic model and its application to the conditions in the cell, *Eur. J. Biochem.*, 45, 39, 1974.
22. **Rose, I. A. and Warms, J. V. B.,** pH dependence of the α-glucose 1,6-diphosphate inhibition of hexokinase II, *Arch. Biochem. Biophys.*, 171, 678, 1975.
23. **Beitner, R., Haberman, S., and Livni, L.,** Complementarity in the regulation of phosphoglucomutase, phosphofructokinase and hexokinase: the role of glucose-1,6-bisphosphate, *Biochim. Biophys. Acta*, 397, 355, 1975.
24. **Rijksen, G. and Staal, G. E. J.,** Regulation of human erythrocyte hexokinase by glucose-1,6-diphosphate and inorganic phosphate, *FEBS Lett.*, 80, 61, 1977.
25. **Rijksen, G. and Staal, G. E. J.,** Regulation of human erythrocyte hexokinase. The influence of glycolytic intermediates and inorganic phosphate, *Biochim. Biophys. Acta*, 485, 75, 1977.
26. **Ponce, J., Roth, S., and Harkness, D. R.,** Kinetic studies on the inhibition of glycolytic kinases of human erythrocytes by 2,3-diphosphoglycerate, *Biochim. Biophys. Acta*, 250, 63, 1971.
27. **Srivastava, S. K. and Beutler, E.,** The effect of normal red cell constituents on the activities of red cell enzymes, *Arch. Biochem. Biophys.*, 148, 249, 1972.

28. **Duhm, J.**, Glycolysis in human erythrocytes containing elevated concentrations of 2,3-P_2-glycerate, *Biochim. Biophys. Acta*, 385, 68, 1975.
29. **Kosow, D. P., Oski, F. A., Warms, J. V. B., and Rose, I. A.**, Regulation of mammalian hexokinase: regulatory differences between isoenzyme I and II, *Arch. Biochem. Biophys.*, 157, 114, 1973.
30. **Wilson, J. E.**, Ligand-induced conformations of rat brain hexokinase: effect of glucose-6-phosphate and inorganic phosphate, *Arch. Biochem. Biophys.*, 159, 543, 1973.
31. **Ellison, W. R., Lueck, J. D., and Fromm, H. J.**, Studies on the kinetics and mechanism of orthophosphate activation of bovine brain hexokinase, *Biochem. Biophys. Res. Commun.*, 57, 1214, 1974.
32. **Ellison, W. R., Lueck, J. D., and Fromm, H. J.**, Studies on the mechanism of orthophosphate regulation of bovine brain hexokinase, *J. Biol. Chem.*, 250, 1864, 1975.
33. **Uyeda, K.**, Phosphofructokinase, *Adv. Enzymol.*, 48, 193, 1979.
34. **Rijksen, G. and Staal, G. E. J.**, Human erythrocyte hexokinase deficiency. Characterization of mutant enzyme with abnormal regulatory properties, *J. Clin. Invest.*, 62, 294, 1978.
35. **Bachelard, H. S. and Goldfarb, P. S. G.**, Adenine nucleotides and magnesium ions in relation to control of mammalian cerebral-cortex hexokinase, *Biochem. J.*, 112, 579, 1969.
36. **Purich, D. L. and Fromm, H. J.**, The kinetics and regulation of rat brain hexokinase, *J. Biol. Chem.*, 246, 3456, 1971.
37. **Rijksen, G. and Staal, G. E. J.**, Kinetics of human erythrocyte hexokinase. Influence of temperature, ATP^{4-} and magnesium ions, *Biochim. Biophys. Acta*, 452, 382, 1976.
38. **Gerber, G., Berger, H., Jänig, G. R., and Rapoport, S. M.**, Interaction of hemoglobin with ions. Quantitative description of the state of magnesium, adenosine 5′-triphosphate, 2,3-biphosphoglycerate, and human hemoglobin under simulated intracellular conditions, *Eur. J. Biochem.*, 38, 563, 1973.
39. **Rose, I. A., Warms, J. V. B., and Kosow, D. P.**, Specificity for the glucose-6-P inhibition site of hexokinase, *Arch. Biochem. Biophys.*, 164, 729, 1974.
40. **Rijksen, G. and Staal, G. E. J.**, Purification and some properties of human erythrocyte hexokinase, *Biochim. Biophys. Acta*, 445, 330, 1976.
41. **Rapoport, T. A., Heinrich, R., Jacobasch, G., and Rapoport, S.**, A linear steady-state treatment of enzymatic chains. A mathematical model of glycolysis of human erythrocytes, *Eur. J. Biochem.*, 42, 107, 1974.
42. **Marshall, W. E., Costello, A. J. R., Henderson, T. D., and Omachi, A.**, Organic phosphate binding to hemoglobin in intact human erythrocytes determined by 31P nuclear magnetic resonance spectroscopy, *Biochim. Biophys. Acta*, 490, 290, 1977.
43. **Baghurst, P. A. and Nichol, L. W.**, The binding of organic phosphates to human methaemoglobin A. Perturbation of the polymerization of proteins by effectors, *Biochim. Biophys. Acta*, 412, 168, 1975.
44. **Haney, D. N. and Bunn, H. F.**, Glycolysation of hemoglobin in vitro: affinity labeling of hemoglobin by glucose-6-phosphate, *Proc. Natl. Acad. Sci. U.S.A.*, 73, 3534, 1976.
45. **Berger, H., Jänig, G. R., Gerber, G., Ruckpaul, K., and Rapoport, S. M.**, Interaction of hemoglobin with ions. Interactions among magnesium, adenosine 5′-triphosphate, 2,3-bisphosphoglycerate, and oxygenated and deoxygenated human haemoglobin under simulated intracellular conditions, *Eur. J. Biochem.*, 38, 553, 1973.
46. **Akkerman, J. W. N., Ebberink, R. H. M., Lips, J. P. M., and Christiaens, G. C. M. L.**, Rapid separation of cytosol and particle fraction of human platelets by digitonin-induced cell damage, *Br. J. Haematol.*, 44, 291, 1980.
47. **Knull, H. R., Taylor, W. F., and Wells, W. W.**, Effects of energy metabolism on in vivo distribution of hexokinase in brain, *J. Biol. Chem.*, 248, 5414, 1973.
48. **Knull, H. R., Taylor, W. F., and Wells, W. W.**, Insulin effects on brain energy metabolism and the related hexokinase distribution, *J. Biol. Chem.*, 249, 6930, 1974.
49. **Wilson, J. E.**, "Ambiquitous" behavior of brain hexokinase. Rapid and reversible interaction of hexokinase with the outermitochondrial membrane, *Biophys. J.*, 37, 18, 1982.
50. **Rijksen, G., Oskam, R., Molthof, C. F. M., Lee-On, S. J., Streefkerk, M., and Staal, G. E. J.**, Hexokinase isoenzymes from anaplastic and differentiated medullary thyroid carcinoma in the rat, *Eur. J. Cancer Clin. Oncol.*, 20, 967, 1984.
51. **Sprengers, E. D., Koenderman, A. H. L., and Staal, G. E. J.**, Mitochondrial and cytosolic hexokinase from rat brain: one and the same enzyme?, *Biochim. Biophys. Acta*, 755, 112, 1983.
52. **Kurakowa, M., Tokuoka, S., Oda, S., Tsubotani, E., and Ishibashi, S.**, Difference in efficiency of function between mitochondrial bound hexokinase and non-bound one, *Biochem. Int.*, 2, 645, 1981.
53. **Bustamante, E. and Pedersen, P. L.**, High aerobic glycolysis of rat hepatoma cells in culture: role of mitochondrial hexokinase, *Proc. Natl. Acad. Sci. U.S.A.*, 74, 3735, 1977.
54. **Ballatori, N. and Cohen, J.**, Intracellular distribution of hexokinase in the tissue zones of rat kidney, *Biochim. Biophys. Acta*, 657, 448, 1981.

55. **Bustamante, E., Morris, H. P., and Pedersen, P. L.,** Energy metabolism of tumor cells. Requirement for a form of hexokinase with a propensity for mitochondrial binding, *J. Biol. Chem.*, 256, 8699, 1981.
56. **Wilson, J. E. and Felgner, P. L.,** An inverse relation between mitochondrial hexokinase content and phosphoglucomutase activity of rat tissues, *Mol. Cell. Biochem.*, 18, 39, 1977.
57. **Lusk, J. A., Manthorpe, C. M., Kao-Jen, J., and Wilson, J. E.,** Predominance of the cytoplasmic form of brain hexokinase in cultured astrocytes, *J. Neurochem.*, 34, 1412, 1980.
58. **Wilson, J. E.** The latent hexokinase activity of rat brain mitochondria, *Biochem. Biophys. Res. Commun.*, 28, 123, 1967.
59. **Thompson, M. F. and Bachelard, H. S.,** Cerebral cortex hexokinase: comparison of properties of solubilized mitochondrial and cytoplasmic activities, *Biochem. J.*, 118, 25, 1970.
60. **Bigl, V., Muller, L., and Bieshold, D. J.,** Some properties of soluble and solubilized particle-bound hexokinase, *J. Neurochem.*, 18, 721, 1971.
61. **Thompson, M. F. and Bachelard, H. S.,** Differences in catalytic properties between cerebral cytoplasmic and mitochondrial hexokinase, *Biochem. J.* 161, 593, 1977.
62. **Bigl, B., Biesold, D., Dowedowa, E. L., and Pigarewa, S. D.,** Die Hexokinase- aktivität verschiedener Hirnregionen der Katze, *Acta Biol. Med. Germ.*, 26, 27, 1971.
63. **Rose, S. A. and Warms, J. V. B.,** Mitochondrial hexokinase. Release, rebinding and location, *J. Biol. Chem.*, 242, 1635, 1967.
64. **Rijksen, G.,** unpublished observations, 1982.
65. **Felgner, P. L., Messer, J. L., and Wilson, J. E.,** Purification of a hexokinase binding protein from the outer mitochondrial membrane, *J. Biol. Chem.*, 254, 4946, 1979.
66. **Ureta, T.,** Phylogeny, ontogeny and properties of hexokinase from vertebrates, in *Isozymes*, Vol. 3, Markert, C. L., Ed., Academic Press, New York, 1975, 575.
67. **Kosow, D. P. and Rose, I. A.,** Ascites tumor mitochondrial hexokinase II. Effect of binding on kinetic properties, *J. Biol. Chem.*, 243, 3630, 1968.
68. **Rose, I. A. and Warms, J. V. B.,** Stability of hexokinase II in vitro and in ascites tumor cells, *Arch. Biochem. Biophys.*, 213, 625, 1982.
69. **Fiek, C., Benz, R., Roos, N., and Brdiczka, D.,** Evidence for identity between the hexokinase-binding protein and the mitochondrial porin in the outer membrane of rat liver mitochondria, *Biochim. Biophys. Acta*, 688, 429, 1982.
70. **Spence, J. T. and Pitot, H. C.,** Hormonal regulation of glucokinase in primary cultures of adult rat hepatocytes, *J. Biol. Chem.*, 254, 12331, 1979.
71. **Malaisse, W. J., Sener, A., and Levy, J.,** The stimulus-secretion coupling of glucose-induced insulin release. Fasting-induced adaption of key glycolytic enzymes in isolated islets, *J. Biol. Chem.*, 251, 1731, 1976.
72. **Niemeyer, H., Ureta, T., and Clark-Turri, L.,** Adaptive character of liver glucokinase, *Mol. Cell. Biochem.*, 6, 109, 1975.
73. **Spence, J. T., Merrill, M. J., and Pitot, H. C.,** Role of insulin, glucose and cyclic GMP in the regulation of glucokinase in cultured hepatocytes, *J. Biol. Chem.*, 256, 1598, 1981.
74. **Belfiore, F., Romeo, F., Napoli, E., and Vecchio, L. L.,** Enzymes of glucose metabolism in liver of subjects with adult onset diabetes, *Diabetes*, 23, 293, 1974.
75. **Katzen, H. M., Soderman, D. D., and Wiley, C. E.,** Multiple forms of hexokinase. Activities associated with subcellular particulate and soluble fractions of normal and streptozotocin diabetic rat tissues, *J. Biol. Chem.*, 245, 4081, 1970.
76. **Bernstein, R. S. and Kipnis, D. M.,** Regulation of rat hexokinase isoenzymes. II. Effects of growth hormone and dexamethasone, *Diabetes*, 22, 923, 1973.
77. **Bernstein, R. S. and Kipnis, D. M.,** Regulation of rat hexokinase isoenzymes. I. Assay and effect of age, fasting and refeeding, *Diabetes*, 22, 913, 1973.
78. **Bernstein, R. S., Marshall, M. C., and Carney, A. L.,** Effects of dietary composition on adipose tissue hexokinase-II and glucose utilization in normal and streptozotocin-diabetic rats, *Diabetes*, 26, 770, 1977.
79. **Belfiore, F., Rabuazzo, A. M., Napoli, E., Borzi, V., and Vecchio, L. L.,** Enzymes of glucose metabolism and of the citrate cleavage pathway in adipose tissue of normal and diabetic subjects, *Diabetes*, 24, 865, 1975.
80. **Kraaijenhagen, R. J., Rijksen, G., and Staal, G. E. J.,** Hexokinase isozyme distribution and regulatory properties in lymphoid cells, *Biochim. Biophys. Acta*, 631, 402, 1980.
81. **Knull, H. R., Taylor, W. F., and Wells, W. W.,** Insulin effects on brain energy metabolism and the related hexokinase distribution, *J. Biol. Chem.*, 249, 6930, 1974.
82. **Anderson, J. W. and King, P.,** Subcellular distribution of hexokinase activity in rat jejunal mucosa: response to diabetes and dietary changes, *Biochem. Med.*, 12, 1, 1975.

83. **Bessman, S. and Geiger, P. J.,** Compartmentation of hexokinase and creatine phosphokinase, cellular regulation and insulin action, in *Current Topics in Cellular Regulation,* Vol. 16, Horecker, B. L. and Stadtman, E. R., Eds., Academic Press, New York, 1980, 55.
84. **Schauer, M., Heinrich, R., and Rapoport, S. M.,** Mathematische modellierung der Glycolyse und des Adeninnukleotidstofwechsels menschlicher Erythrozyten. I. Reaktionskinetische Ansätze, Analyse des in vivo-Zustandes und Bestimmung der Anfangsbedingungen für die in vitro-Experimenter, *Acta Biol. Med. Germ.,* 40, 1659, 1981.
85. **Valentine, W. N., Oski, F. A., Paglia, D. E., Baughan, M. A., Schneider, A. S., and Naiman, J. L.,** Hereditary hemolytic anemia with hexokinase deficiency. Role of hexokinase in erythrocyte aging, *N. Engl. J. Med.,* 276, 1, 1967.
86. **Paglia, D. E., Shende, A., Lanzkowsky, P., and Valentine, W. N.,** Hexokinase "New Hide Park": a low activity erythrocyte isozyme in a chinese kindred, *Am. J. Hematol.,* 10, 107, 1981.
87. **Gilsanz, F., Meyer, E., Paglia, D. E., and Valentine, W. N.,** Congenital hemolytic anemia due to hexokinase deficiency, *Am. J. Dis. Child.,* 132, 636, 1978.
88. **Beutler, E., Dyment, P. G., and Masumoto, F.,** Hereditary nonspherocytic hemolytic anemia and hexokinase deficiency, *Blood,* 51, 935, 1978.
89. **Keitt, A. S.,** Hemolytic anemia with impaired hexokinase activity, *J. Clin. Invest,* 48, 1997, 1969.
90. **Necheles, T. F., Rai, U. S., and Cameron, D.,** Congenital nonspherocytic hemolytic anemia associated with an unusual erythrocyte hexokinase abnormality, *J. Lab. Clin. Med.,* 76, 593, 1970.
91. **Moser, D., Ciresa, M., and Schwarzmeier, J.,** Hexokinasemangel bei hämolytischer Anämie, *Med. Welt,* 21, 1977, 1970.
92. **Siimes, M. A., Rahalia, E.-L., and Leisti, J.,** Hexokinase deficiency in erythrocytes: a new variant in 5 members of a Finnish family, *Scan. J. Haematol.,* 22, 214, 1979.
93. **Board, P. G., Trueworthy, R., Smith, J. E., and Moore, K.,** Congenital nonspherocytic hemolytic anemia with an unstable hexokinase variant, *Blood,* 51, 111, 1978.
94. **Newman, P., Muir, A., and Parker, A. C.,** Non-spherocytic haemolytic anemia in mother and son associated with hexokinase deficiency, *Br. J. Haematol.,* 46, 537, 1980.
95. **Rijksen, G., Akkerman, J. W. N., van den Wall Bake, A. W. L., Pott Hofstede, D., and Staal, G. E. J.,** Generalized hexokinase deficiency in the blood cells of a patient with non-spherocytic hemolytic anemia, *Blood,* 61, 12, 1983.
96. **Harris, J. W. and Kellermeyer, R. W.,** *The Red Cell,* 3rd ed., Commonwealth Fund and Harvard University Press, Cambridge, 1974, 464.
97. **Anderson, J. W., Herman, R. H., Tyrell, J. B., and Cohn, R. M.,** Hexokinase: a compartmented enzyme, *Am. J. Clin. Nutr.,* 24, 642, 1971.
98. **Löhr, G. W., Waller, H. D., Anschütz, F., and Knopp, A.,** Hexokinasemangel in Blutzellen bei einer Sippe mit familiärer Panmyelopathie (typ Fanconi), *Klin. Wochenschr.,* 43, 870, 1965.
99. **Goebel, K. M,, Gassel, W. D., Goebel, F. D., and Kaffarnik, H.,** Hemolytic anemia and hexokinase deficiency associated with malformations, *Klin. Wochenschr.,* 50, 849, 1972.
100. **Akkerman, J. W. N.,** Regulation of carbohydrate metabolism in platelets. A Review, *Thromb. Haemostas.,* 39, 712, 1978.
101. **Kester, M. V., Phillips, T. L., and Gracy, R. W.,** Changes in glycolytic enzyme levels and isozyme expression in human lymphocytes during blast transformation, *Arch. Biochem. Biophys.,* 183, 700, 1977.
102. **Rogers, P. A., Brenton, D. P., and Hopkinson, D. A.,** Changes in the activity and isozyme patterns of glycolytic enzymes during stimulation of normal human lymphocytes with phytohaemagglutinin, *Ann. Hum. Genet. London,* 42, 213, 1980.
103. **Akkerman, J. W N., Rijksen, G., Gorter, G., and Staal, G. E. J.,** Platelet functions and energy metabolism in a patient with hexokinase deficiency, *Blood,* 63, 147, 1984.
104. **Boulard, M. R., De Bellis, R., Blume, K. G., Sorel, M., and Staal, G. E. J.,** Influence of copper on the activity of human red cell hexokinase, *I.R.C.S.,* 2, 1049, 1074.
105. **De Bellis, R., Boulard, M. R., Kasdorf, H., Rodriquez, I., Ferrando, R., Dilandro, J., Ferrari, M., Sanguinett, C. M., and Tanzer, J.,** Metabolic changes in red blood cells in malignant lymphomas, *Br. J. Haematol.,* 42, 35, 1979.
106. **Bearn, A. G.,** Wilson's disease, in *The Metabolic Basis of Inherited Disease,* 3rd ed., Stanbury, J. B., Wijngaarden, J. B., and Fredrickson, D. S., Eds., McGraw-Hill, New York, 1972, chap. 43.
107. **Warburg, O.,** *The Metabolism of Tumors,* Constable, London, 1930.
108. **Sato, S., Matsushima, T., and Sugimura, T.,** Hexokinase isozyme patterns of experimental hepatomas of rat, *Cancer Res.,* 29, 1437, 1969.
109. **Farron, F.,** The isoenzymes of hexokinase in normal and neoplastic tissues of the rat, *Enzyme,* 13, 233, 1972.
110. **Bustamante, E. and Pedersen, P. L.,** Mitochondrial hexokinase of rat hepatoma cells in culture: solubilization and kinetic properties, *Biochemistry,* 19, 4972, 1980.

111. **Emmelot, P. and Bos, C. J.,** Differences in the association of two glycolytic enzymes with plasma membranes isolated from rat liver and hepatoma, *Biochim. Biophys. Acta*, 121, 434, 1966.
112. **Dadidova, S. Y., Shapot, V. S., and Solowjewa, A. A.,** Hexokinase activity and glycolytic capacity of plasma membranes of hepatomas, *Biochim. Biophys. Acta*, 158, 303, 1968.
113. **Sydow, G.,** Uber Unterschiede in der Aktivität und intrazellular Localisation der Hexokinase zwischen normaler Rattenleber und Hepatomen, *Acta Biol. Med. Germ.*, 20, 1, 1968.
114. **Kikuchi, Y., Sato, S., and Sugimura, T.,** Hexokinase isozyme patterns of human uterine tumors, *Cancer*, 30, 444, 1972.
115. **Sekiya, S., Kikuchi, Y., and Takamizawa, H.,** High- and low-tumorigenic culture lines of rat uterine adenocarcinoma and their isozyme patterns of lactate dehydrogenase and hexokinase, *Cancer Res.*, 33, 3324, 1973.
116. **Kamel, R. and Schwarzfischer, F.,** Hexokinase isozymes in human neoplastic and fetal tissues: the existence of hexokinase II in malignant tumors and in placenta, *Humangenetik*, 30, 181, 1975.
117. **Bennett, M. J., Timperley, W. R., Taylor, C. B., and Hill, A. S.,** Isozymes of hexokinase in the developing, normal and neoplastic human brain, *Eur. J. Cancer*, 14, 189, 1978.
118. **Sprengers, E. D., van Veelen, C. W. M., Rijksen, G., and Staal, G. E. J.,** Hexokinase from tumour cells, presented at the 6th Meeting of the European Association for Cancer Research, Budapest, October 12 to 15, 1981, 221.
119. **Beemer, F. A., Vlug, A. M. C., Rousseau-Merck, M. F., van Veelen, C. W. M., Rijksen, G., and Staal, G. E. J.,** Glycolytic enzymes from human neuroectodermal tumors of childhood, *Eur. J. Cancer Clin. Oncol.*, 20, 253, 1984.
120. **Beemer, F. A., Vlug, A. M. C., Rijksen, G., Hamburg, A., and Staal, G. E. J.,** Characterization of some glycolytic enzymes from human retina and retinoblastoma, *Cancer Res.*, 42, 4228, 1982.
121. **Suda, M., Tanaka, T., Yanagi, S., Hayashi, S., Imamura, K., and Taniuchi, K.,** Dedifferentiation of enzymes in the liver of tumor-bearing animals, *GANN Monogr. Cancer Res.*, 13, 79, 1972.
122. **Herzfeld, A., Greengard, O., and McDermott, W.,** Enzyme pathology of the liver in patients with and without nonhepatic cancer, *Cancer*, 45, 2383, 1980.
123. **Racker, E.,** Why do tumor cells have a high aerobic glycolysis?, *J. Cell. Physiol.*, 89, 697, 1979.
124. **Johnson, J. H., Zimniak, A., and Racker, E.,** Inhibition of hexokinase and protein kinase activity of tumor cells by a chloromethyl ketone derivative of lactic acid, *Biochemistry*, 21, 2984, 1982.
125. **Stevens, V. J., Vlassara, H., Abati, A., and Cerami, A.,** Non-enzymatic glycosylation of hemoglobin, *J. Biol. Chem.*, 252, 2998, 1977.
126. **Tegos, C. and Beutler, E.,** Red cell glycolytic intermediates in diabetic patients, *J.Lab. Clin. Med.*, 96, 85, 1980.
127. **Strickland, J. M. and Ellis, D. A.,** Isozymes of hexokinase in human muscular dystrophy, *Nature (London)*, 253, 464, 1975.
128. **Strickland, J. M. and Ellis, D. A.,** Isoenzymes of human skeletal muscle hexokinase in muscular dystrophy, *Biochem. Soc. Trans.*, 2, 1125, 1974.
129. **Ellis, D. A., Strickland, J. M., and Tye, J. E.,** Human hexokinase isoenzymes of muscle in health and disease, *Biochem. Soc. Trans.*, 4, 1054, 1976.
130. **Beitner, R. and Nordenberg, J.,** The regulatory role of glucose-1,6-diphosphate in muscle of dystrophic mice, *FEBS Lett.*, 98, 199, 1979.
131. **Beitner, R., Haberman, S., Nordenberg, J., and Cohen, T. J.,** The levels of cyclic GMP and glucose-1,6-diphosphate, and the activity of phosphofructokinase, in muscle from normal and dystrophic mice, *Biochim. Biophys. Acta*, 542, 537, 1978.
132. **Ohga, K., Sakuraba, H., Suzuki, Y., and Kobayashi, N.,** Biochemical abnormalities in the Wiscott-Aldrich syndrome: low activity of hexokinase in lymphocytes, *Acta Paediatr. Jpn.*, 23, 72, 1981.
133. **Akkerman, J. W. N. and Rijksen, G.,** unpublished results, 1983.
134. **Ueda, M., Taketa, K., Shimamura, J., and Kosaka, K.,** Hepatic hexokinase isozyme in liver diseases, *Clin. Chim. Acta*, 53, 381, 1974.
135. **Aoe, H.,** Histochemical studies of hepatic hexokinase and glucose 6-phosphatase in chronic liver diseases, *Acta Histochem. Cytochem.*, 8, 120, 1975.
136. **Ueda, M., Taketa, K., and Kosaka, K.,** Hexokinase isozyme pattern in CCl_4-injured rat liver, *Clin. Chim. Acta*, 60, 77, 1975.
137. **Bielicki, L. and Krieglstein, J.,** Solubilization of brain mitochondrial hexokinase by thiopental, *Arch. Pharmacol.*, 298, 61, 1977.
138. **Höfeler, H. and Krieglstein, J.,** Solubilization of hexokinase activity by an effect of thiopental on the mitochondrial membrane, *Pharmacology*, 24, 156, 1982.

Chapter 5

PHOSPHORYLATION OF PHOSPHOFRUCTOKINASE — THE POSSIBLE ROLE OF COVALENT MODIFICATION IN THE REGULATION OF GLYCOLYSIS

Hans Werner Hofer

TABLE OF CONTENTS

I. INTRODUCTION

A. Phosphofructokinase and the Regulation of Glycolysis

Fructose-1,6-bisphosphate was the first phosphorylated intermediate of glycolysis whose existence was detected.[1] The enzymatic reaction leading to its formation was formulated 30 years later as follows:

$$\text{Fructose-6-phosphate} + \text{ATP} \rightarrow \text{Fructose-1,6-bisphosphate} + \text{ADP} + \text{H}^+ \qquad (1)$$

by Ostern et al.[2] The enzyme catalyzing the reaction is now commonly called phosphofructokinase.

Broad interest was evoked when it was realized that phosphofructokinase was responsible for the Pasteur effect and catalyzed one of the irreversible and rate-limiting steps of glycolysis in the cells.[3-7] Most enzymes of glycolysis maintain the concentrations of the partners of their reactions close to thermodynamic equilibrium. This, however, is not true for the reactants and products of Reaction 1, as is illustrated by the data summarized in Table 1. The free energies of glycolytic reactions were calculated on the basis of metabolite concentrations determined in the extract of mouse diaphragms after 1 hr incubation in the presence of glucose. As originally pointed out by Bücher,[8] the glycolytic reactions can be grouped into two different categories by their free energies under *in situ* conditions. To the first group belong the reactions whose partners are in pseudo-equilibrium, and the second group is formed by nonequilibrium reactions. Bücher's idea becomes more transparent when the data of Table 1 are plotted as a flow diagram (Figure 1). With the exception of phosphofructokinase and pyruvate kinase, the glycolytic enzymes in the diaphragms obviously were sufficiently active to establish concentrations of the reaction partners close to the thermodynamic equilibrium. Phosphofructokinase and pyruvate kinase exhibited enzyme activities, when measured under optimum reaction conditions in the in vitro assay, in the same order of magnitude as the activities of enzymes catalyzing equilibrium reactions, e.g., aldolase. In spite of this fact, they were incapable of equilibrating reactants and products of their reactions. In the case of phosphofructokinase, incomplete equilibration of reactants and products of the reaction was not primarily due to the distinctly negative $\Delta G_o'$ of the reaction. This was underlined by the fact that a quasi-equilibrium was attained in the phosphoglycerate kinase reaction whose free energy under standard conditions is still more negative. Several other reasons were also responsible for the nonequilibrium of the phosphofructokinase reaction:

1. The limited supply of fructose-6-phosphate. The rate of hexosemonophosphate formation in the diaphragm, as in many tissues, is controlled by glucose uptake into the cell and by the regulation of glycogen degradation.
2. The high ratio of ATP to ADP concentrations. The maintenance of this ratio is one of the primary aims of metabolic control and is guaranteed by the high capacity of ATP synthetizing metabolic systems.
3. The allosteric properties of phosphofructokinase in the presence of physiological pH values in combination with the inhibiting influence of negative effectors, e.g., ATP.

As was clearly shown for perfused muscle,[9,10] the maximum rates of glycolysis observed were only 1 to 5% of phosphofructokinase activity as measured in vitro. The data suggested that phosphofructokinase was never completely deinhibited. The strong inhibition of the potential activity of phosphofructokinase is obviously to maintain the potency for regulation. Activation of the enzyme under exergonic reaction conditions will increase the rate of fructose-1,6-bisphosphate formation and hence of glycolysis.

Table 1
FREE ENERGIES OF GLYCOLYTIC REACTIONS UNDER *IN SITU* CONDITIONS

Reaction catalyzed by	ΔG (kJ)
Phosphoglucose isomerase	−1.7
Phosphofructokinase	−14.7
Fructose-1,6-bisphosphate aldolase	+0.4
Glyceraldehydephosphate dehydrogenase + phosphoglycerate kinase	−0.8
Phosphoglycerate mutase	−0.4
Enolase	+1.2
Pyruvate kinase	−34.8
Lactate dehydrogenase	−0.8

Note: Freshly excised hemi-diaphragms of rats were incubated in the presence of 5.0 m*M* glucose in Krebs-Henseleit bicarbonate-buffer gassed with 95% O_2 and 5% CO_2 (pH 7.4). Free energies were calculated on the basis of the concentrations of metabolites and cofactors determined after 1 hr of incubation (37°C). An intracellular pH of 7.0 was assumed. If ΔG^{o} values for 37°C were not available, they were calculated from the equilibrium constants. The conversion of the equilibrium constant K_1 (at a given absolute temperature T_1) to the value K_2 at temperature T_2 was performed by the formula $\ln(K_2/K_1) = \Delta H \cdot (T_2 - T_1)/R \cdot T_1 \cdot T_2$ (R = gas constant).

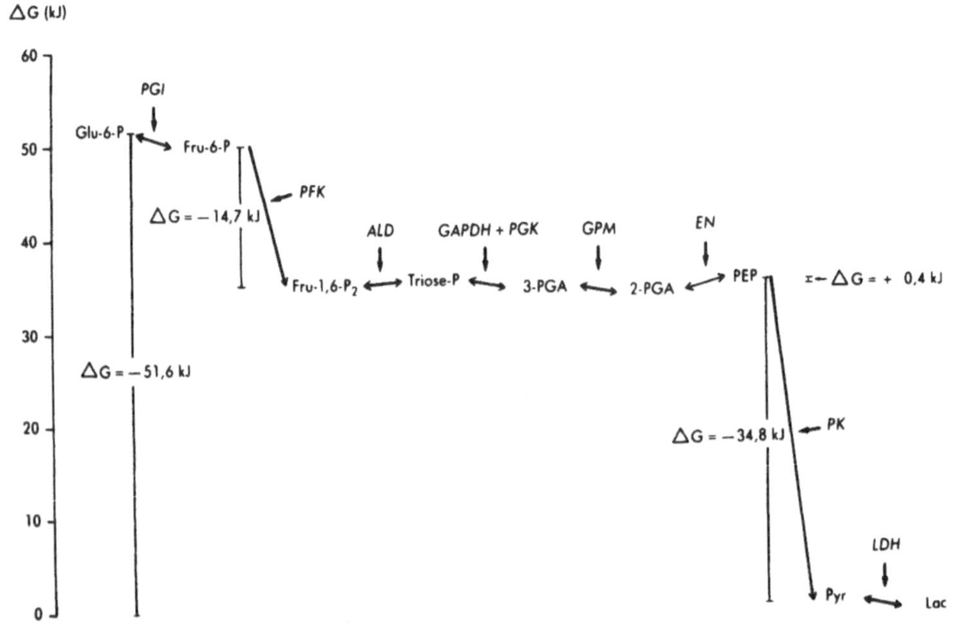

FIGURE 1. Flow diagram of nonequilibrium and quasi-equilibrium reactions of glycolysis under *in situ* conditions.

B. Mechanisms Influencing Phosphofructokinase Activity

The study of the regulatory kinetics of phosphofructokinase was initiated by Lardy and Parks[11] when they discovered the inhibition of the enzyme by ATP. Later, the cooperative binding of fructose-6-phosphate and the deinhibition of phosphofructokinase by positive effectors were described for the mammalian enzyme by Lowry and Passonneau[12] and for the yeast enzyme by Vinuela et al.[13] In the following years, the regulatory kinetics of phosphofructokinase were thoroughly studied by several groups, and various activators and inhibitors of the enzyme were detected. In 1980, Sols[14] had compiled a list of 26 different effectors of phosphofructokinase. Since then, fructose-2,6-bisphosphate was detected, a substance which is regarded to be one of the most powerful regulators of phosphofructokinase.

Although the regulation of phosphofructokinase could be easily explained by the action of a very limited number of effectors and mechanisms, the factual control of the enzyme activity appears to be more complex and to imply also other mechanisms in addition to the allosteric control. Hofer and Pette[15] pointed out that phosphofructokinase was inactivated by dissociation and that this inactivation was prevented by the presence of allosteric effectors. The cooperativity of the fructose-6-phosphate saturation curve and also the $K_{0.5}$ values were significantly affected by the concentration of phosphofructokinase in the assays, and this characteristic of the enzyme was referred to the association state of the enzyme.[16,17] Cross-linking of the enzyme by bifunctional reagents led to the stabilization of the reduced sensitivity of the enzyme against inhibition by ATP.[18-20]

Another factor severely affecting phosphofructokinase activity is the oxidation state of a highly reactive SH-group.[21-24] The oxidation of the cysteine residue, and thus the inactivation of the enzyme, is retarded by the presence of ATP and also of positive effectors. It was recently postulated that there could be a general mechanism which could control the redox state of SH-groups also in vivo.[25]

Finally, it was demonstrated that phosphofructokinase from mammals and also from the parasite *Ascaris suum* could be reversibly phosphorylated, and that these enzymes occurred in a partially phosphorylated state also in vivo.[26-29] With regard to the well-known mechanisms of enzyme regulation by energy-dependent interconversion, and especially by reversible phosphorylation (cf., References 30 and 31) it was assumed that phosphorylation was also of functional significance in the case of phosphofructokinase.

The structural, kinetic, and regulatory properties of phosphofructokinase have been excellently reviewed by several authors.[32-37] Therefore, we will restrict this discussion to a survey of the effects of phosphorylation on phosphofructokinase. Since many aspects of the phosphorylation of phosphofructokinase from liver have been recently reviewed by Söling and Brand,[38] emphasis will be put on phosphofructokinase from skeletal muscle.

II. EVIDENCE FOR INTERCONVERSION OF PHOSPHOFRUCTOKINASE

A. Evidence for Different States of Phosphofructokinase Activity

In 1962 Mansour and Mansour[39] reported that the addition of serotonine elevated the activity of phosphofructokinase when it was added to a homogenate prepared from liver flukes (*Fasciola hepatica*). Since a stimulation of cAMP production by serotonine had been previously observed in the fluke,[40] the authors supposed that the action of serotonine on phosphofructokinase was mediated by cAMP. The activation of phosphofructokinase by cAMP was found to be exerted in two different ways. First, the cyclic nucleotide activated the ATP inhibited enzyme. This effect was related to the allosteric properties of phosphofructokinase. Second, phosphofructokinase was also markedly activated when the concentrated enzyme was preincubated with cAMP, in spite of the fact that the nucleotide was diluted in the assays below the level required for allosteric activation.

Effects of cAMP and other adenine nucleotides, indicating a conversion of different states of activity of phosphofructokinase into each other, were also observed by several other authors. Vinuela et al.[41] reported that incubation of crude yeast extracts, in the presence of fluoride, led to the disappearance of the marked sensitivity of the enzyme to ATP inhibition. This occurred also when dialyzed extracts were incubated in the presence of ATP-Mg^{2+} and cAMP. The parallel between the activation of yeast phosphofructokinase and of phosphorylase b suggested that a phosphorylation reaction was involved. The studies on yeast phosphofructokinase were later extended by Holzer and co-workers.[42-45] They demonstrated that a protein factor was involved in the desensitization reaction. It was also shown that the protein factor could be replaced by ADP and that an energy-dependent interconversion of phosphofructokinase was not involved in the desensitization process.

Other systems providing evidence for the presence of different active states of phosphofructokinase were liver and heart muscle. Ramaiah and co-workers[46,47] observed the fixation of purified liver phosphofructokinase in a high affinity state for fructose-6-phosphate when the enzyme was incubated in the presence of ATP, inorganic phosphate, NH_4^+, K^+, and fluoride. Mansour[48] found significant activation of heart muscle phosphofructokinase after incubation of the enzyme in the presence of ATP-Mg^{2+}. Yet, it was not possible to detect substantial phosphorylation of the enzyme. More recent studies of the perfused rat heart have revealed that phosphofructokinase activity as measured in extracts of heart muscle was stimulated when epinephrine was added to the perfusions.[49-51] The increased activity was stable even after the removal of low-molecular-weight substances by gel filtration and was referred to a diminished sensitivity of phosphofructokinase towards inhibition by citrate and ATP. The epinephrine action was probably mediated by adrenergic alpha-receptors. The authors, however, found no evidence that a phosphorylation-dephosphorylation mechanism was involved.

An enhancement of the specific activity of phosphofructokinase in homogenates from quiescent cultures of 3T3 cells was observed by Rozengurt et al.[52] after the addition of serum, epidermal growth factor, or insulin to the cultures. The increased cellular phosphofructokinase activity was concomitant with a stimulated flux rate of glycolysis.[53] The increase of the phosphofructokinase activity was not affected by the addition of cycloheximide to the cultures nor by gel filtration of the extracts. However, evidence for the involvement of enzyme phosphorylation in phosphofructokinase activation was not presented. Further, it remained unclear whether the stimulation of phosphofructokinase activity was an isolated phenomenon or was accompanied by the increase of the activities of other glycolytic enzymes.

B. Evidence for the Modification of Phosphofructokinase by Phosphorylation

Evidence for the incorporation of phosphate into phosphofructokinase was first presented by Brand and Söling[26] in 1975. Phosphofructokinase, present in a protein fraction which was precipitated from a rat liver extract at pH 5.5, was inactivated by incubation in the presence of Mg^{2+}. On the other hand, the phosphofructokinase activity was elevated when the fraction was incubated in the presence of ATP and Mg^{2+}. The rate of reactivation of previously inactivated phosphofructokinase depended on the pretreatment of the rats from which the extracts had been prepared. The rate was low when the rats had been starved and increased again when the rats had been refed after starvation.

Inactive and active fractions of phosphofructokinase were separated by chromatography of the pH 5.5 precipitates on DEAE-cellulose. A protein peak containing inactive phosphofructokinase was eluted at a phosphate concentration of less than 100 m*M*. The phosphofructokinase of this peak was reactivated by incubation in the presence of ATP-Mg^{2+}. Catalytically active phosphofructokinase was eluted by 120 m*M* phosphate. The enzyme was inactivated by the addition of a protein fraction that was eluted from the column by high salt concentrations (about 0.5 *M* NaCl). These experiments gave rise to the assumption

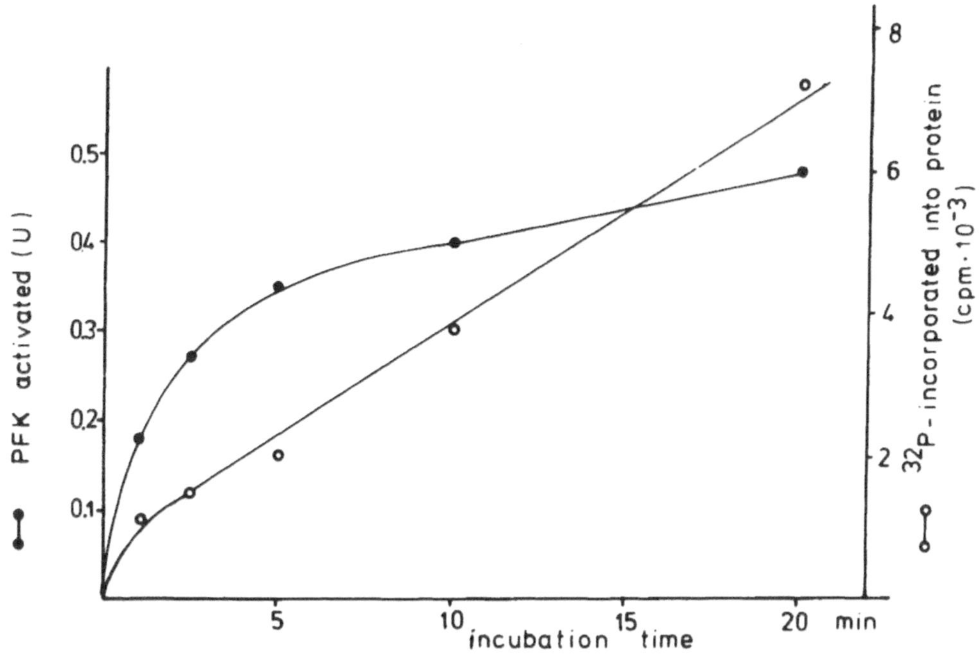

FIGURE 2. Time dependence of the activation of phosphofructokinase in the presence of [γ-^{32}P]ATP – Mg^{2+} (closed circles) and of the incorporation of radioactivity into the protein precipitated by antibodies against liver phosphofructokinase (open circles). (From Brand, I. A. and Söling, H. D., *FEBS Lett.*, 57, 163, 1975. With permission.)

that the inactive phosphofructokinase was associated with an activating enzyme, presumably a kinase, and that phosphofructokinase inactivation was due to the action of a phosphatase.

Brand and Söling[26] also reported the incorporation of radioactive phosphate into phosphofructokinase when the inactive enzyme was incubated in the presence of [γ-^{32}P]ATP. Radioactivity was found in the protein pellet when the phosphofructokinase was subsequently precipitated by antibodies (cf., Figure 2). However, the activity of phosphofructokinase and the [^{32}P]phosphate content in the antibody precipitate did not increase in parallel. Incorporation of [^{32}P]phosphate into the protein proceeded almost linear with time during a period of 20 min whereas phosphofructokinase reactivation occurred preferentially during the first 5 min of incubation. These experiments did not definitely prove that the reactivation of phosphofructokinase was the consequence of phosphofructokinase phosphorylation, but they have considerably stimulated the research on phosphofructokinase regulation in liver.

Phosphorylation of phosphofructokinase under in vivo conditions was first demonstrated by Hofer and Fürst.[27] Injection of radioactive [^{32}P]phosphate into mice led to the incorporation of radioactivity into phosphofructokinase, demonstrated by the autoradiographs of sodium dodecyl sulfate electrophoreses of purified phosphofructokinase. Based on the specific radioactivity of the γ-phosphate of ATP in the tissue extracts, the extent of phosphorylation was calculated at about one phosphate per phosphofructokinase tetramer.

The demonstration of phosphorylation of phosphofructokinase under in vitro and in vivo conditions inaugurated a number of further studies which will be reviewed here with emphasis on mammalian systems. It should be noted, however, that phosphofructokinase phosphorylation is not restricted to phosphofructokinase from mammals but was also demonstrated to occur in the nematode *Ascaris suum*.

III. ENZYMES PHOSPHORYLATING AND DEPHOSPHORYLATING PHOSPHOFRUCTOKINASE

A. Phosphorylation of Phosphofructokinase

1. cAMP-Dependent Protein Kinases

The phosphorylation of glycolytic enzymes and other proteins by cAMP-dependent protein kinase was studied by Cohen et al.[54] The highest initial rate of phosphorylation was obtained with the β-subunit of phosphorylase kinase as a substrate. Setting this rate as 100%, rates higher than 10% were observed with glycogen synthetase (80%), L-type pyruvate kinase (35%), protein phosphatase inhibitor-1 (30%), and the α-subunit of phosphorylase kinase (20%). The rate of phosphorylation of phosphofructokinase relative to the rate of the phosphorylation of the β-subunit of phosphorylase kinase was reported to be only 0.1%. This rate, however, was still considerably higher than the rate of phosphorylation of other glycolytic enzymes.

Kirschenlohr and Hofer[55] have determined the extent of phosphorylation of glycolytic enzymes, mainly from rabbit muscle, by the catalytic subunit of the cAMP-dependent protein kinase. They observed that after 1 hr incubation, there was 5.8 times more radioactive phosphate incorporated into phosphorylase kinase than into phosphofructokinase. But they also found that under identical reaction conditions no other glycolytic enzyme contained more than 5% of the amount of phosphate bound to phosphofructokinase.

These data clearly indicated the phosphorylating potency of cAMP-dependent protein kinase for phosphofructokinase. On the other hand, phosphofructokinase exhibited much poorer qualities as a substrate for the protein kinase than enzymes like phosphorylase kinase or glycogen synthetase which have been shown to be substrates of cAMP-dependent protein kinase also under physiological conditions. However, the data provided by Cohen et al.[54] and Kirschenlohr and Hofer[55] presumably underestimated the true values of phosphorylation of phosphofructokinase. In the absence of its second substrate, fructose-6-phosphate, phosphofructokinase acts as an ATPase with about 0.05 to 1.0% of its fructose-6-phosphate phosphorylating activity. In the presence of micromolar concentrations of the enzyme and less than 1 m*M* ATP, i.e., conditions that were used for the protein phosphorylation studies, the ATP concentrations must have considerably decreased during the incubation period.[57,58]

An extensive study on the ability of catalytic subunit of the cAMP-dependent protein kinase to phosphorylate phosphofructokinase was performed by Kemp and co-workers.[59] At pH 7.0, the maximum velocity of phosphofructokinase phosphorylation was 44% of that of histone phosphorylation. Based on that value, 1 mg of purified catalytic subunit of cAMP-dependent protein kinase should have incorporated 100 nmol phosphate into phosphofructokinase per minute. However, the Michaelis constant, was very high and amounted to 230 nmol/mℓ, corresponding to a phosphofructokinase concentration of 18.4 mg/mℓ. This value is about 40 times higher than the average concentration of phosphofructokinase in rabbit muscle. Therefore, the actual phosphofructokinase phosphorylating activity of catalytic subunit of the cAMP-dependent protein kinase in tissue should be limited to about 2.5% of the maximum activity. Assuming similar properties of skeletal muscle cAMP-dependent protein kinase as those of heart muscle protein kinase used in the study of Kemp et al.,[59] the fraction of phosphofructokinase that can be phosphorylated within 1 min by cAMP-dependent protein kinase is less than 5% of total phosphofructokinase.

Kemp et al.[59] have also studied the influence of allosteric regulators of phosphofructokinase on its phosphorylation. The rate of phosphorylation was increased by 34% in the presence of 1 m*M* 5′-AMP and by 91% in the presence of 0.1 m*M* fructose-1,6-bisphosphate. On the other hand, the phosphorylation rate was decreased to 65% of the control value in the presence of 2 m*M* citrate. Although there was an obvious influence of phosphofructokinase conformation on the ability of cAMP-dependent protein kinase to phosphorylate phosphofructokinase, none of the effectors changed the Michaelis constant of the protein kinase for phosphofructokinase.

The studies of Kemp et al.[59] clearly showed that phosphorylation of phosphofructokinase by cAMP-dependent protein kinase in vivo must be a slow process, at least when its rate is compared to the phosphorylation of the enzymes involved in glycogen metabolism. Yet, it cannot be ruled out that in the future effectors will be found which may be responsible for turning phosphofructokinase to a better substrate of cAMP-dependent protein kinase.

2. Ca^{2+}-Dependent Protein Kinases

In a preliminary study, Hofer and Soerensen-Ziganke[60] have demonstrated the ability of phosphorylase kinase to phosphorylate phosphofructokinase. In the presence of 1 nmol phosphofructokinase per milliliter, the initial activity of phosphorylase kinase was about 1% of that obtained with phosphorylase b as a substrate. This would imply that the phosphorylating capacity of phosphorylase kinase for phosphofructokinase in muscle has an upper limit of 25 nmol/min·g muscle. In spite of the fact that phosphofructokinase is only a poor substrate for phosphorylase kinase, as compared to glycogen phosphorylase, the relatively high concentrations of this protein kinase in muscle could provide a higher phosphorylating potency for phosphofructokinase than cAMP-dependent protein kinase.

B. Dephosphorylation of Phosphofructokinase

1. In Vitro Dephosphorylation by Intestinal Phosphatase

Foe and Kemp[61] showed that rabbit muscle phosphofructokinase which had been phosphorylated either in vivo or in vitro was effectively dephosphorylated by alkaline phosphatase from bovine intestine. Yet, the specific activity of the phosphatase in the presence of 1 mg phosphofructokinase per milliliter was 10^5-fold lower than against 5 m*M* nitrophenyl phosphate. EDTA, dithioerythritol, and arsenate inhibited the phosphorylation reaction, while Mg^{2+} ions activated it. Phosphofructokinase, labeled either in vitro or in vivo, was completely dephosphorylated in the presence of 10 m*M* $MgCl_2$. The rates of dephosphorylation were not significantly different for phosphofructokinase labeled by different phosphorylation modes. Dephospho-phosphofructokinase was rephosphorylated by the catalytic subunit of cAMP-dependent protein kinase. Foe and Kemp also clearly demonstrated that the solubilization of bound radioactive phosphate by intestinal phosphatase was not a proteolytic effect but led to the release of inorganic phosphate. Different extents of [^{32}P] incorporation into native and dephosphorylated phosphofructokinase were not observed, in spite of the fact that native phosphofructokinase already contained about 0.2 mol phosphate per protomer. The discrepancy in the extent of phosphorylation of native and dephosphorylated phosphofructokinase could indicate some conformational change in phosphofructokinase that would limit in vitro phosphorylation to a less than stoichiometric extent.

2. Dephosphorylating Enzyme Activities for Phosphofructokinase in Muscle

[^{32}P]-Phosphofructokinase, labeled by cAMP-dependent protein kinase, was dephosphorylated by an extract from rabbit muscle.[60] The initial rate of dephosphorylation was 0.14 nmol/min·g muscle. The concentration of phosphofructokinase in the same muscle was calculated as 0.7 nmol/g muscle. The phosphatase activity appeared to be low compared to phosphofructokinase concentration, yet the reaction conditions were not optimized. However, it was established that the product of the dephosphorylation was inorganic phosphate.

Phosphorylase phosphatase preparations prepared by ethanol treatment were active against phosphorylase *a* but exhibited only low dephosphorylating activity against phosphofructokinase. A sample of protein phosphatase which was prepared by the method of Varsanyi and Heilmeyer,[63] however, dephosphorylated phosphofructokinase with a considerable rate. The initial rate of phosphofructokinase dephosphorylation was 24% of the dephosphorylating activity for phosphorylase *a*.[60] However, the method of Varsanyi and Heilmeyer[63] did not include purification steps leading to the dissociation of high-molecular-weight protein phosphatase forms to low-molecular-weight forms.

IV. THE NATURE OF THE PHOSPHORYLATED SITES

A. The Nature of the Phosphorylated Amino Acid

1. Isolation of [^{32}P]-Labeled Phosphofructokinase

In principle, two possibilities exist to introduce labeled phosphate into phosphofructokinase: (1) by the administration of [^{32}P]phosphate to animals and incorporation into the enzyme in vivo, and (2) by in vitro labeling of the enzyme in the presence of a protein kinase and [γ-^{32}P]ATP.

Both methods have been successfully applied by several laboratories. The in vivo method guaranteed high specificity and physiological significance but suffered from the great amounts of radioactive material that had to be used and the low specific radioactivities which were finally present in the purified enzyme. The in vitro labeling procedures were much more convenient and led to phosphofructokinase preparations containing [^{32}P] of high specific radioactivity. However, it was necessary to check whether the same sites of phosphofructokinase were phosphorylated in vitro and in vivo.

A number of attempts have been made to separate the radioactive phosphate, which had been incorporated into phosphofructokinase in vivo, from the protein. This was necessary because the labeled phosphate might possibly have been a constituent of nucleic acids purified together with the enzyme[123] or of effectors like ATP or fructose-1,6-bisphosphate, which are tightly bound to phosphofructokinase. In our laboratory, the radioactivity which coprecipitated with the enzyme in 10% trichloroacetic acid, was regarded to be covalently bound to the protein when it remained in the precipitate after several washings.

Figure 3 shows the normal pattern of sodium dodecylsulfate electrophoresis of mice muscle phosphofructokinase after staining with Coomassie Blue (a). The molecular weight of the subunits was the same as that of rabbit muscle phosphofructokinase (molecular weight 90,000). Figure 3b represents the corresponding autoradiography of the gel. It is obvious that the pattern on the film and the pattern of staining are identical.

2. The Phosphorylated Amino Acid Residue

Figure 4 shows the results of the paper electrophoreses of the acid hydrolysate of rabbit muscle phosphofructokinase which had been labeled with [^{32}P] in vivo. The majority of the amino acids migrated to the cathode at pH 2.8; phosphoserine migrated to the anode. The same migration distance was observed for the labeled amino acid from phosphofructokinase by autoradiography. Inorganic phosphate migrated about twice as fast.[72,75]

A similar analysis was performed by Brand and Söling[26] with phosphofructokinase from liver. They also found phosphoserine as the phosphorylated amino acid. Presently, there exists no evidence that another amino acid besides serine is phosphorylated in phosphofructokinase either in vivo or in vitro.

3. Isolation of the Phosphorylated Peptides

In order to investigate the nature of the phosphorylated sites of phosphofructokinase, Soerensen-Ziganke and Hofer[72] isolated the radioactive peptides after the labeling of the enzyme with [^{32}P]phosphate in vivo and after incubation of the purified enzyme with cAMP-dependent protein kinase in the presence of [γ-^{32}P]ATP in vitro.

Phosphofructokinase preparations, labeled by different methods, were split into peptides by CNBr treatment and trypsin digestion. Chromatography of the peptides on Bio-Gel® P-

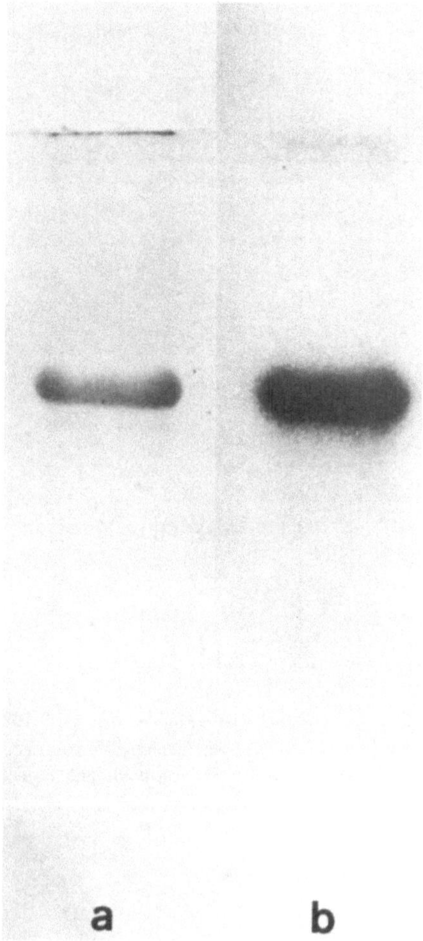

FIGURE 3. Electrophoresis of [^{32}P]phosphofructokinase from mice muscle on polyacrylamide gels in the presence of sodium dodecylsulfate. (a) Staining of the gel with Coomassie Blue. (b) Autoradiography of the gel.

30 and Sephadex® G-25 revealed no difference between the molecular weights of the radioactive peptides labeled by different modes. The chromatography fractions containing the radioactive peptides were subject to peptide mapping on thin-layer plates. The positions of the radioactive peptides were visualized by autoradiography as shown in Figure 5A and 5B. Two radioactive spots were detected on each plate. They were distinctly separated in the chromatography and migrated in opposite directions in the subsequent electrophoreses which were performed at pH 6.1. The chromatographic and electrophoretic mobilities of the two peptides were essentially the same for the peptides labeled either in vivo or by cAMP-dependent protein kinase in vitro.

B. Limited Proteolysis of Phosphofructokinase

1. Release of a C-Terminal Phosphorylated Peptide by the Action of Proteases

Riquelme and Kemp[70] reported that a brief exposure of labeled phosphofructokinase to subtilisin converted nearly all radioactivity into a form that was soluble in 5% trichloroacetic acid, while the bulk of the enzyme protein was precipitated. The loss in enzyme activity,

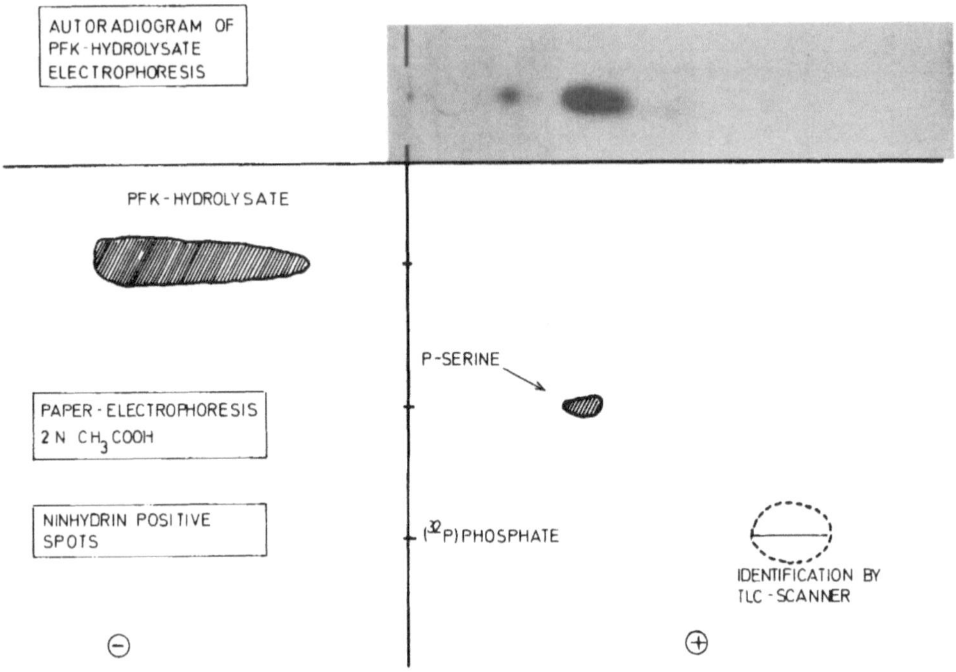

FIGURE 4. Paper electrophoresis of the hydrolysate of an in vivo labeled [^{32}P]phosphofructokinase. The lower part of the figure shows the positions of amino acids visualized by ninhydrin staining and the position of [^{32}P]phosphate. The upper part represents the pattern on the X-ray film after autoradiography of the electrophoresis of the phosphofructokinase hydrolysate indicating the position of the labeled amino acid.

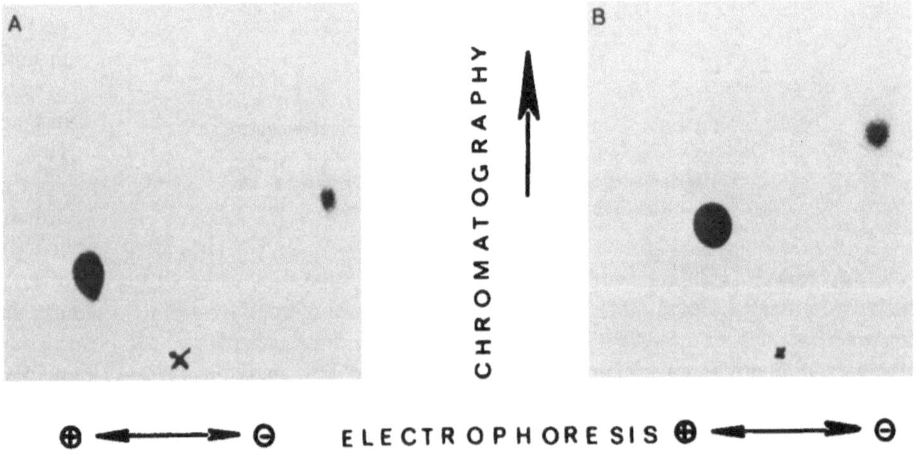

FIGURE 5. Autoradiographes of the peptide maps obtained from the radioactivity-containing peptides from in vivo (A) and in vitro (B) labeled [^{32}P]Phosphofructokinase. (From Soerensen-Ziganke, B. and Hofer, H. W., *Biochem. Biophys. Res. Commun.*, 90, 204, 1979. With permission.)

as measured before trichloroacetic acid precipitation, did not parallel radioactivity release. Electrophoresis of the precipitable protein moiety provided an explanation of this phenomenon. After a 2 min incubation of phosphofructokinase, about 30% of radioactive phosphate was released. However, the catalytic activity of the enzyme was still complete and the

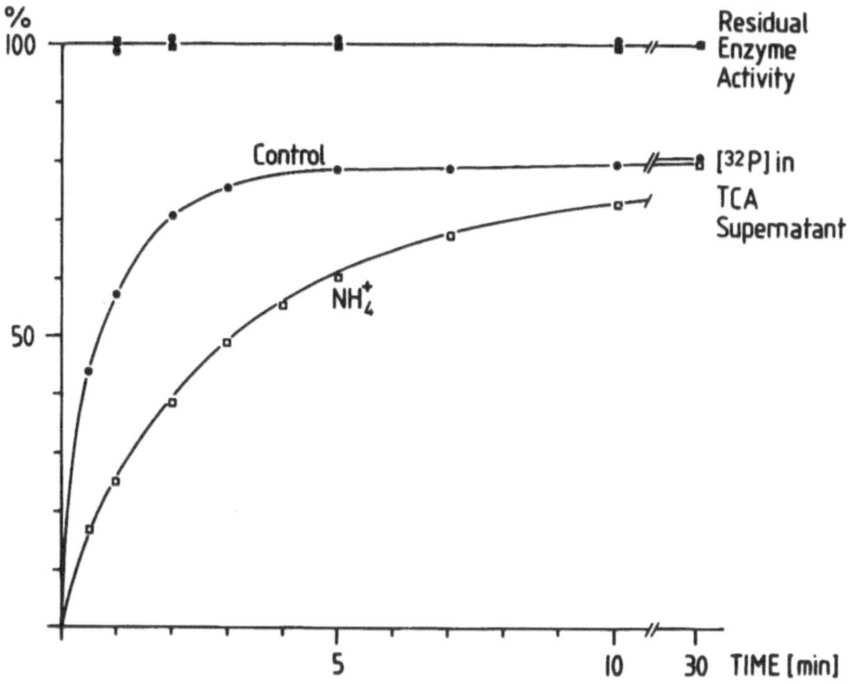

FIGURE 6. Time dependence of the release of radioactivity from [32P]phosphofructokinase by treatment with trypsin in the absence and presence of NH_4^+ ions. (From Krystek, E. and Hofer, H. W., *Biochem. Biophys. Res. Commun.*, 99, 1138, 1981. With permission.)

electrophoresis of the trichloroacetic acid precipitates on sodium dodecylsulfate gels showed the same pattern as in the absence of subtilisin. Longer incubations resulted in the partial loss of enzyme activity, which coincided with the appearance of a 72,000-Dalton unit of phosphofructokinase.

A similar phenomenon was observed in our laboratory[74] when we used trypsin instead of subtilisin. Trypsin led to a rapid release of trichloroacetic acid-soluble radioactivity from [32P]-labeled phosphofructokinase (Figure 6). In the presence of 0.002 mg trypsin per milligram phosphofructokinase, one half of the radioactivity was released within less than 30 sec. About 80% of the radioactivity was solubilized within 5 min. The residual radioactivity remained insoluble in trichloroacetic acid even after prolonged incubation. However, the catalytic activity of phosphofructokinase was not affected by incubation of the enzyme in the presence of low concentrations of trypsin for more than 1 hr. Figure 7 shows sodium dodecylsulfate electrophoreses on polyacrylamide gels of the trichloroacetic acid precipitates of phosphofructokinase incubated with trypsin for different periods of time. During the first 5 min, i.e., the period during which most of the radioactivity became soluble in trichloroacetic acid, the protein consisted almost entirely of a single band with the approximate molecular weight 90,000. After prolonged incubation, cleavage into smaller fragments took place. The electrophoresis patterns then showed a predominant fragment with about 40,000 mol wt. This fragment appeared to correspond to that observed by Emerk and Frieden.[76] These authors also reported that the activity of phosphofructokinase, under optimum assay conditions, was not influenced by that cleavage. Yet, before a substantial amount of the enzyme was split into the 40,000 mol wt fragments, another proteolytic process took place which had escaped detection in the study of Emerk and Frieden. A basic peptide was released from the enzyme to which 80% of the radioactive label was bound after the incubation of the

FIGURE 7. Electrophoreses of the trichloroacetic acid precipitates of phosphofructokinase incubated for different periods of time in the presence of trypsin. Polyacrylamide gel electrophoreses were performed in the presence of 0.1% sodium dodecylsulfate. The samples, precipitated 0 min (a), 2 min (b), 5 min (c), 20 min (e), and 30 min (f) after the addition of trypsin with trichloroacetic acid, were obtained in the control experiment shown in Figure 6. (From Krystek, E. and Hofer, H. W., *Biochem. Biophys. Res. Commun.*, 99, 1138, 1981. With permission.)

enzyme with [γ-^{32}P]ATP in the presence of catalytic subunit of cAMP-dependent protein kinase. This peptide had an approximate molecular weight of less than 1500, and therefore, its release did not significantly change the molecular weight of the phosphofructokinase protomers. Due to its basic character, the peptide migrated to the cathode on sodium dodecylsulfate electrophoresis, and since it did not contain aromatic amino acids, it was not monitored by its absorbance. The peptide was supposedly located at, or close to, one of the termini of the peptide chain of phosphofructokinase and seemed to be connected to the remainder of the peptide chain by an extremely labile peptide bond.

The peptide containing the major fraction of radioactivity, obtained by the trypsin treatment, could be removed from the active moiety of phosphofructokinase by gel filtration. Obviously, the peptide was not required for catalytic activity and was not tightly bound to the active part of the enzyme after the cleavage of the peptide bond. This is in contrast to the strong forces between the 40,000 mol wt fragments of phosphofructokinase which were obtained by longer incubation with trypsin. These products of proteolysis stick together during gel filtration.

2. *Influence of Substrates and Effectors on the Release of the Phosphate-Containing Peptide*

As shown by Figure 6, the rate of the tryptic release of the phosphate-containing peptide was decreased by the presence of NH_4^+ ions. The initial rate was reduced to 29% of the control. K^+ ions decreased the rate to 51%. ATP, fructose-6-phosphate, and fructose-1,6-bisphosphate were significantly less effective compared to NH_4^+ and K^+. The decrease of the rate of tryptic release of the phosphopeptide by these substances was only 13%.

Riquelme and Kemp[70] also reported an influence of substrates and effectors on the reaction of phosphofructokinase with subtilisin. They found a strong inhibition of phosphopeptide release and of the loss of enzyme activity by ATP and $MgCl_2$. The presence of $MgCl_2$ and citrate strongly inhibited the cleavage of the phosphopeptide. $(NH_4)_2SO_4$ partly inhibited the decrease of catalytic activity of phosphofructokinase without affecting the release of the phosphopeptide.

C. Amino Acid Sequence of the Phosphorylated Site

The amino terminus of phosphofructokinase, as determined by dansylation and high pressure liquid chromatography after hydrolysis, is phenylalanine.[151] The same amino acid was found at the NH_2-terminus after mild trypsin treatment of phosphofructokinase. The carboxyl terminus released by carboxypeptidase Y was valine. Short treatment of the enzyme with a low trypsin concentration changed the COOH-terminus to arginine. The labeled phosphopeptides obtained by treatment of phosphofructokinase with subtilisin and trypsin, were subjected to automatic Edman degradation by Kemp and co-workers.[59] They found the phosphoserine in a tryptic octapeptide, whose sequence was also present in the dodecapeptide that was released by subtilisin. The sequence of the C-terminal phosphopeptide, as determined by Kemp et al.,[59] is

(Glx-Leu)-His-Ile-Ser-Arg-Lys-Arg-Ser(P)-Gly-Glu-Ala-Thr-Val·COOH

/ Tryptic peptide /

/ Subtilisin peptide /

Determination of the COOH-terminal amino acid after the release of the subtilisin peptide by carboxypeptidase Y led to the conclusion that the sequence continues in the direction of the amino-terminus with leucine and glutamic acid or glutamine.[151]

The C-terminal site of phosphofructokinase, as determined by Kemp et al.[59] has common characteristics with the phosphorylation site of other proteins. A sequence of

Arg-Arg-Ala-Ser

of a synthetic peptide was found to be an optimum substrate for cAMP-dependent protein kinase by Kemp et al.[77,78] They also demonstrated that the replacement of the alanine next to serine by an arginine and the addition of an alanine on the C-terminal side of the serine considerably reduced the maximum velocity of cAMP-dependent protein kinase with the synthetic substrate. This fact may explain the low phosphorylation rate of cAMP-dependent protein kinase with phosphofructokinase as a substrate as compared to the rate of phosphorylation of phosphorylase kinase.

D. Additional Phosphorylation Sites

About 80 to 90% of the radioactivity attached to phosphofructokinase by cAMP-dependent protein kinase in the presence of [γ-^{32}P]ATP and about 80% of the radioactivity introduced into phosphofructokinase in vivo were released by short-term treatment of the enzyme in the presence of low concentrations of trypsin. The 10 to 20% of radioactivity remaining in the trichloroacetic acid precipitate were only reduced after prolonged treatment of phosphofructokinase with the protease. Such a treatment also reduced enzyme activity. This observation may suggest the presence of a second phosphorylated site on the enzyme and would be in agreement with the fact that a minor peak of radioactivity was routinely found in chromatographies of tryptic digests of phosphofructokinase when analyzed by thin-layer or high pressure liquid chromatography. The amino acid analyses of both phosphorylated peptides gave similar results. Taking into account that only very specific sequences of amino acids serve as substrates for cAMP-dependent protein kinase, the similarity of the amino acid composition is not an irrefutable argument for the identity of both phosphorylation sites.

There were also arguments for a single phosphorylation site on phosphofructokinase. In contrast to Soerensen-Ziganke and Hofer,[72] Uyeda et al.[73] and Kemp et al.[59] observed only one radioactive spot on tryptic fingerprints of labeled phosphofructokinase. Whether the low radioactivity on the second site prevented its detection, or whether incomplete tryptic cleavage in the experiments of Soerensen-Ziganke and Hofer was responsible for the discrepancy cannot be decided at the present. However, it should be noted that two phosphorylation sites were also found in the phosphofructokinase from *Ascaris suum*. In contrast to rabbit muscle phosphofructokinase, these sites were phosphorylated to equal extent. In a recent study, Brand et al.[147] presented fingerprints of tryptic peptides of phosphofructokinase from rat liver which also showed more than one phosphorylated peptide.

V. THE EXTENT OF PHOSPHOFRUCTOKINASE PHOSPHORYLATION IN VIVO

A. The Phosphate Content of Phosphofructokinase and Other Phosphoproteins in Muscle

The original finding of Hofer and Furst,[27] that phosphofructokinase isolated from skeletal muscle contained covalently bound phosphate, has been confirmed by several laboratories for the enzyme from skeletal muscle,[68-70,73,85] liver,[38,65] and erythrocytes.[66] It was also demonstrated that phosphofructokinase from liver,[38,147] kidney,[67] and erythrocytes[66] was phosphorylated by cAMP-dependent protein kinase. In erythrocytes which contain both L- and M-subunits of phosphofructokinase,[79] radioactive phosphate was incorporated into both types of subunits to the same extent.[66]

1. Kinetics of Incorporation of [^{32}P]Phosphate into Phosphofructokinase

The kinetics of the incorporation of [^{32}P]phosphate intraperitoneally injected into mice are shown in Figure 8. The level of the inorganic labeled phosphate in the serum rapidly reached a maximum. The following decline was heterophasic due to the exchange of phosphate in various tissues. The specific radioactivity of [^{32}P]phosphate in muscle reached a maximum 20 hr after injection. The same was true for the specific radioactivity of the gamma-phosphate of ATP and of the radioactivity of phosphofructokinase. However, the

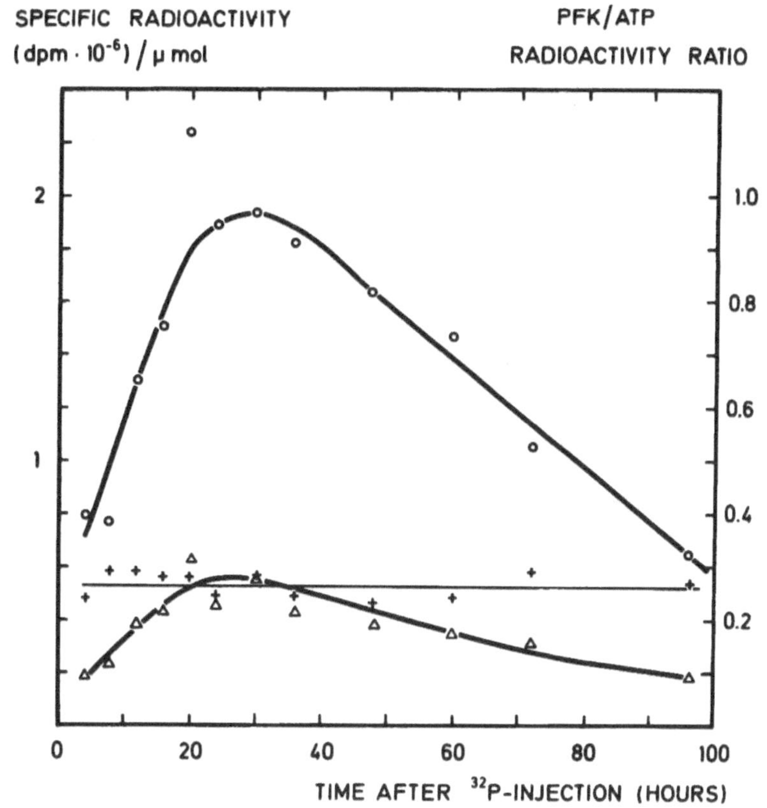

FIGURE 8. Time course of the specific radioactivities of the γ-phosphate group of ATP (○ – ○) and of phosphofructokinase (△ – △) from mice muscle after the injection of [^{32}P]phosphate. (From Hofer, H. W. and Soerensen-Ziganke, B., *Biochem. Biophys. Res. Commun.*, 90, 199, 1979. With permission.)

specific radioactivities were lower, especially in the case of phosphofructokinase. Forty hours after phosphate injection, a significant difference between the specific radioactivities of phosphate and of ATP was no longer observed. The ratios of the specific radioactivities of phosphofructokinase and ATP remained constant at approximately 0.25, independent of time. Thus, on the average, every fourth subunit of the phosphofructokinase tetramer was phosphorylated.

2. *Inhibition of Protein Synthesis and Phosphorylation of Phosphofructokinase*

The constant ratio between the specific radioactivities of phosphofructokinase and ATP, as apparent from Figure 9, indicated a rapid equilibration of phosphate between ATP and phosphofructokinase. This suggested that the phosphorylation of phosphofructokinase was a reversible short-term process. However, taking into account that the half-life of phosphofructokinase in muscle is in the range of 7 to 17 hr,[134,135] the data of Figure 9 did not rule out the possibility of an irreversible modification of phosphofructokinase. Arguments against that possibility were contributed by studies on the incorporation of [^{32} P]phosphate in the presence of inhibitors of protein synthesis.[64,75]

The incorporation of [1-^{14} C]leucine into mice muscle phosphofructokinase was reduced to 19% of the controls when the animals were treated with cycloheximide (cf., Table 2). Protein synthesis was inhibited to that extent over a period of 21 hr, when repeated doses of 0.015 mg of cycloheximide per gram body weight were applied in 3-hr intervals. In spite

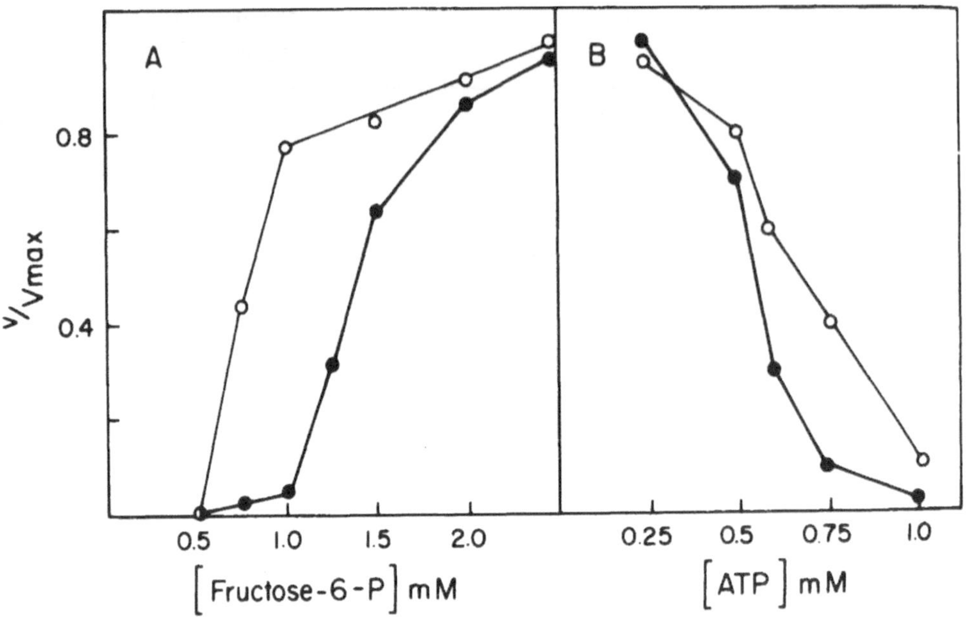

FIGURE 9. Comparison of kinetics of native (open symbols) and in vitro phosphorylated (closed symbols) phosphofructokinase. (From Foe, L. G. and Kemp, R. G., *J. Biol. Chem.*, 257, 6368, 1982. With permission.)

Table 2
INCORPORATION OF [1-^{14}C]LEUCINE AND [^{32}P]PHOSPHATE INTO MICE MUSCLE PHOSPHOFRUCTOKINASE IN THE ABSENCE AND PRESENCE OF CYCLOHEXIMIDE

	PFK-bound radioactivity	
Condition	[^{14}C] dpm/mol	[^{32}P] dpm/mol
Control	220,080 ± 33,230	138,940 ± 28,390
Cycloheximide[a]	41,520 ± 14,740	123,060 ± 48,280
Cycloheximide/control	0.19	0.89

[a] Six doses of 0.4 mg cycloheximide/g body weight in intervals of 3 hr.

Modified from Hofer, H. W. and Soerensen-Liganke, B., *Biochem. Biophys. Res. Commun.*, 90, 199, 1979.

of inactivation of protein synthesis, the phosphofructokinase activity in muscle remained constant. Obviously, the inhibition of protein synthesis also prevented the degradation of the enzyme. However, the incorporation of radioactive phosphate into phosphofructokinase was not affected by cycloheximide. Thus, the experiment ruled out that phosphofructokinase phosphorylation was connected with the process of protein synthesis; otherwise the incorporation of the radioactivity of [1-^{14}C]leucine and [^{32}P]phosphate should have been inhibited to the same degree. On the other hand, phosphorylation was not enhanced by impairment of protein degradation. However an increase of [^{32}P] incorporation would be expected, if phosphofructokinase phosphorylation was an irreversible process.

Table 3
SOLUBLE PHOSPHOPROTEINS FROM MICE MUSCLE[55,56]

	Relative value of			
Mol wt	Radioactivity	Protein	Specific radioactivity	Phosphorylation
220,000	50	80	0.6	1.2
130,000	1	0.6	1.7	2.0
115,000	2	2	1.0	1.1
97,000[a]	100	100	1.0	1.0
90,000[b]	12	37	0.3	0.28
64,000	30	4	7.5	5.0
58,000	4	1	4.0	2.4
49,000	10	3	3.3	1.5
49,000	9	2	4.5	2.5
49,000	60	20	3.0	1.5
48,000	2	0.8	2.5	1.5
43,000	20	3	6.7	3.1
41,000	1	2	0.5	0.2
33,000	3	9	0.3	0.1
25,000	30	4	7.5	2.1
21,000	2	2	1.0	0.2
17,500	20	2	10.0	1.8

[a] Identified as phosphorylase a.
[b] Identified as phosphofructokinase.

3. Degree of Phosphorylation of Phosphofructokinase Relative to Other Phosphoproteins

Another study which used the incorporation of radioactive [^{32}P]phosphate into proteins, as a tracer for protein phosphorylation in muscle, was performed by Kirschenlohr and Hofer.[56] They evaluated the patterns obtained from muscle extracts by two-dimensional electrophoresis for protein and radioactivity using densitometric methods. More than 30 spots were visible on the autoradiographs. A partial list is given in Table 3 arranged according to the molecular weight as estimated from the mobility in sodium dodecylsulfate electrophoresis. Radioactive components, whose protein character was questionable, were not included in Table 3. Protein concentration and radioactivity content of glycogen phosphorylase were both set to unity. The relative amounts of protein and radioactivity for the individual spots were listed in columns 2 and 3 of Table 3 together with the relative specific radioactivities depicted in column 4.

The relative specific radioactivities, as defined by the ratio of relative radioactivity and relative protein content, ranged from 0.2 to 6.6. Since the relative radioactivity as depicted in column 4 of Table 3 is not directly related to the phosphorylation of the individual proteins on a molar basis, an "index of phosphorylation" was included in column 5 of the table. It was defined by the product of the relative specific radioactivity and the molecular weight of the protein divided by the molecular weight of phosphorylase. This index corrected the specific radioactivities by the molecular weight of the protein subunits. The indices calculated for the phosphoproteins varied between 0.2 and 5, but values between 1.0 and 2.5 occurred most frequently. Thus, most of the phosphoprotein subunits contained about the same amount or twice the amount of phosphate as present in the subunit of phosphorylase *a* which contains a single phosphorylation site.

4. Phosphorylation of Glycolytic Enzymes in Muscle

The incorporation of radioactive phosphate into other glycolytic enzymes was also studied by Kirschenlohr and Hofer.[55,56] The positions of the glycolytic enzymes on the two-dimensional electrophoresis of crude muscle extracts were established by comparison with the electrophoretic pattern obtained for purified enzymes. With the exception of phosphofructokinase, no other radioactive component listed in Table 3 corresponded to a glycolytic enzyme. This finding was corroborated by partial purification of the glycolytic enzymes from the muscles of mice which had been treated with [^{32}P]phosphate. Subsequent analysis performed as described above, led to the detection of small amounts of radioactivity in phosphoglucomutase and phosphoglycerate mutase. However, the radioactivity content in these enzymes was less than 5% of that present in phosphofructokinase, and it was released when the mutases were dialyzed against their unlabeled substrates. Therefore, the authors concluded that the radioactivity in the partially purified mutases represented phosphorylated serine residues within the catalytic center of these enzymes. As was demonstrated by Naijar and Pullman,[81] these serine residues are phosphorylated during the catalytic cycle of the mutase reactions.

B. Influence of the Metabolic State on the Phosphorylation of Phosphofructokinase

1. Influence of Muscle Contraction and Epinephrine — Analysis of Phosphate Content with Chemical Methods

Several laboratories determined the phosphate contents of phosphofructokinase isolated from rabbit skeletal muscle by chemical methods. In contrast to the studies which used the incorporation of radioactive phosphate into phosphofructokinase as a measure for the phosphorylation state of the enzyme, the determination of the phosphate content by chemical methods was not compromised by the possibility of incomplete equilibration of radioactivity between the γ-phosphate group of ATP and phosphofructokinase. When dephosphorylation during the purification procedure was not prevented by special precautions, numbers between 0.5 and 2 phosphate residues per phosphofructokinase tetramer were found.[64,73,75,85,129] Since dephosphorylation of phosphofructokinase in postmortal tissues proceeded very rapidly,[64] these values did not really reflect the phosphorylation state of the enzyme in vivo.

Soerensen-Ziganke and Hofer[64] showed that the phosphorylation state of phosphofructokinase markedly depended on the functional conditions. The phosphate content of purified rabbit muscle phosphofructokinases, obtained in a more extended series of experiments, are summarized in Table 4. It should be noted that the method of phosphate determination was different from that used in earlier experiments.[64] This fact partly explains that the values in Table 4 are lower than those reported before.

The values presented in Table 4 are characteristic for phosphofructokinase from gastrocnemius muscle of anesthesized rabbits. The muscles were excised after freeze-clamping, and the purification of the enzyme was performed in the presence of EDTA and fluoride in order to prevent dephosphorylation.

As listed in Table 4, the phosphate content of phosphofructokinase from nonstimulated muscle was 2.16 mol phosphate per tetramer. The standard deviation of about 25% suggested that there was some variation from animal to animal even within the same condition of treatment. Contraction increased the phosphate content significantly to 3.85 mol phosphate per phosphofructokinase tetramer. The effect of isoproterenol was more complex. Within 30 sec after the injection of beta-adrenergic agonists, the phosphate content of phosphofructokinase markedly increased when compared to that of the resting muscle. However, the increase of phosphorylation was transient. Muscle samples taken 2 min after the injection of isoproterenol contained phosphofructokinase, whose phosphorylation grade was lower than that of the enzyme from the resting control muscle.

The changes in the phosphorylation state of phosphofructokinase from muscle appeared to be very rapid. The rate of these changes was obviously in the same order of magnitude as the interconversion processes in which the glycogenolytic enzymes are involved. The rapid changes of the phosphate content of phosphofructokinase following the injection of

Table 4
PHOSPHATE CONTENT OF PHOSPHOFRUCTOKINASE FROM RABBIT MUSCLE

Functional state	mol Phosphate/mol PFK[a]
Resting muscle	2.16 ± 0.56 (n = 9)
Contracting muscle[b]	3.85 ± 0.53 (n = 8)[d]
Resting muscle + epinephrine[c]	1.67 ± 0.33 (n = 3)

[a] Phosphofructokinase tetramer (mol wt 360,000) from the gastrocnemius muscle.
[b] Electrostimulation (5 impulses/sec) for 2 min.
[c] Single dose (0.2 mg/kg) i.v. 2 min before freeze-clamping of the gastrocnemius muscle.
[d] $p<0.005$ compared with controls.

isoproterenol were possibly related to the fact that high doses of the beta-adrenergic stimulator were administered. In spite of the decrease in the phosphate content of phosphofructokinase during prolonged action of the hormone, there was some stimulation of phosphofructokinase phosphorylation in response to a short-time exposure of the muscle to the beta-agonist. However, the increase in phosphorylation grade was more pronounced in response to phasic contraction of the muscle. This may suggest that muscle phosphofructokinase can be phosphorylated as well by Ca^{2+}- as by cAMP-dependent protein kinases under physiological conditions.

2. The Possible Influence of Insulin on the Phosphate Content of Muscle Phosphofructokinase

Insulin was reported to decrease the activity of cAMP-dependent protein kinase.[130] This observation raised the question whether or not insulin deficiency, under diabetic conditions, would stimulate protein phosphorylation. Bazaes et al.[129] studied the phosphate content of phosphofructokinase from genetically diabetic (db/db) mice in relation to normal (db/m) controls. Based on the incorporation of radioactive phosphate, the phosphate content of phosphofructokinase promoters in the normal mice muscle was determined as 0.14 ± 0.03 mol/mol. The phosphofructokinase promoters from the diabetic mice muscle contained 0.27 ± 0.11 mol phosphate per mole, indicating enhanced phosphorylation of the enzyme. The specific activity of phosphofructokinase in the diabetic muscle was about 30% lower than in normal controls. A partial separation of phospho- and dephospho- forms of phosphofructokinase was obtained by anion-exchange chromatography. However, the two forms exhibited only minor differences in their allosteric kinetics. Therefore, the enhanced phosphorylation of phosphofructokinase in the muscle of diabetic mice did not explain the reduced glycolytic rate in the organ.

3. Influences of Hormones on the Phosphorylation of Liver Phosphofructokinase

The influence of different hormones on the phosphorylation of phosphofructokinase from perfused liver was investigated by Kagimoto and Uyeda[28] using the incorporation of [^{32}P]phosphate as a measure of the degree of phosphorylation. The specific radioactivity of ATP was nearly constant in all experiments reported. Epinephrine, insulin, and 20 m*M* glucose exerted no significant influence on phosphofructokinase phosphorylation. Only glucagon stimulated [^{32}P] incorporation into phosphofructokinase, and this enhancement was more than 3.5-fold. The cAMP level was simultaneously increased to an approximately four- to fivefold concentration, compared to the control experiments performed in the absence of hormones and glucose.

The intracellular concentrations of cAMP were also increased by the addition of 0.1 m*M* cAMP to the perfusion medium. Under these conditions, the incorporation of [^{32}P] into phosphofructokinase was stimulated to almost the same extent as in the presence of glucagon. These experiments suggested that phosphorylation of phosphofructokinase in liver involved a beta-receptor mediated cAMP-dependent mechanism.

Stimulation of the incorporation of radioactive phosphate into phosphofructokinase by glucagon was also observed by Söling and Brand.[38,71] Glucagon concentrations of $3 \cdot 10^{-7}$ *M* as well as 10^{-4} *M* dibutyryl cAMP led to a significant increase of the radioactivity content of phosphofructokinase.

4. Effect of Glucose on the Phosphorylation of Liver Phosphofructokinase

In contrast to the results reported by Kagimoto and Uyeda,[28] Brand and Söling[71] observed an enhancement of phosphofructokinase phosphorylation by glucose.

The basal rate of incorporation of phosphate into phosphofructokinase of hepatocytes from starved rats was 0.2 ± 0.1 mol per tetramer. The phosphorylation of the enzyme was reduced when the hepatocytes were incubated in the presence of high lactate or L-alanine concentrations. Addition of 20 m*M* glucose to the incubations increased phosphofructokinase phosphorylation 2.5-fold in hepatocytes from starved rats and 1.3-fold in hepatocytes from fed rats. The phosphate content in the hepatocytes from fed rats was then 0.86 ± 0.22 mol per mol tetramer, i.e., almost every protomer of the enzyme then contained one phosphate residue. L-Alanine abolished the effect of glucose on phosphofructokinase phosphorylation in hepatocytes. Glucagon had no additive effect on phosphofructokinase phosphorylation in hepatocytes isolated from fed rats. However, the hormone stimulated the phosphorylation of L-type pyruvate kinase, whereas glucose did not affect the phosphorylation of this enzyme. Also dibutyryl cAMP did not significantly enhance the phosphorylation of phosphofructokinase in hepatocytes from fed rats above the level obtained with glucose alone. However, the [^{32}P] incorporation into phosphofructokinase from hepatocytes isolated from starved rats was increased to the maximal value by dibutyryl cAMP.

In a recent study, Brand et al.[147] compared the peptide maps derived from phosphofructokinase that was phosphorylated in the presence of glucose or glucagon. In vivo stimulation of the incorporation of labeled phosphate into phosphofructokinase resulted in identical phosphopeptide maps. The positions of the phosphopeptides were identical to those observed for the phosphopeptides labeled by the action of the catalytic subunit of cAMP-dependent protein kinase. Therefore, the authors concluded that in the intact liver cells the phosphofructokinase was phosphorylated by that protein kinase but that the state of phosphorylation was modified by metabolites interacting with phosphofructokinase. The experiments indicated that the degree of phosphorylation of rat liver phosphofructokinase is significantly affected by the metabolic state. Functional significance of phosphofructokinase phosphorylation for metabolic regulation should therefore be expected. However, the mechanism by which phosphorylation could control phosphofructokinase activity was not evident.

VI. EFFECTS OF PHOSPHORYLATION ON THE PROPERTIES OF PHOSPHOFRUCTOKINASE

A. Effects of Phosphorylation on the Physical Properties of Phosphofructokinase

The role of phosphorylation of phosphofructokinase for the function of the enzyme under physiological conditions has remained unclear. The phosphorylation of liver phosphofructokinase was detected by Brand and Söling[26] when they studied the phenomenon of reversible inactivation of the enzyme in the presence of Mg^{2+} ions and the reactivation of crude enzyme preparations by ATP-Mg^{2+}. However, the inactivation of phosphofructokinase could be also explained by the dissociation of the enzyme into inactive dimers and their reassociation to active tetramers due to the associating effect of ATP.[17,19,82,83] This explanation was supported by the fact that Brand et al.[84] were able to separate active and inactive species of phosphofructokinase on a Sepharose 6B column. The elution volume of the inactive phosphofructokinase form was significantly larger than that of the active phosphofructokinase. This suggested different states of aggregation of active and inactive phosphofructokinase species. Gel filtration of purified rabbit muscle phosphofructokinase led to analogous results when performed under similar conditions.[152] It was demonstrated in these experiments that the aggregation of the enzyme in the presence of ATP was not accompanied by phosphorylation.

However, there remained the possibility that phosphorylation of phosphofructokinase favored the associated active state. Hussey et al.[85] reported that separation of phosphofructokinase forms which were phosphorylated to different extent by gel filtration on agarose columns. The leading fractions of the elution peak of phosphofructokinase contained 0.05 to 0.2 mol of alkali-labile phosphate per mole of phosphofructokinase subunit. The trailing fractions had between 0.1 and 0.4 mol of phosphate bound per mole of phosphofructokinase subunit. These experiments suggested a reduced tendency of the higher phosphorylated phosphofructokinase to aggregate. However, the aggregation was not related to changes in the specific catalytic activity of phosphofructokinase.

Uyeda et al.[73] separated rabbit muscle phosphofructokinase by DEAE-cellulose chromatography into two fractions. Phosphofructokinase eluted with 0.1 *M* Tris-phosphate buffer contained 0.31 mol phosphate per subunit, whereas another fraction of phosphofructokinase that was eluted by 0.3 *M* Tris-phosphate had a phosphate content of 0.8 mol per mol subunit. The two fractions of phosphofructokinase differed from each other, not only in their phosphate content, but also with respect to their aggregation state. Sedimentation velocity patterns of the lower phosphorylated phosphofructokinase showed two major peaks exhibiting sedimentation constants of 12 S and 18 S and a minor 24 S peak. The higher phosphorylated phosphofructokinase consisted mainly of 30 S material and minor 24 S, 18 S, and 12 S peaks. Upon 10-fold dilution of the protein concentration, the 24 S and the 18 S phosphofructokinase dissociated into the 12 S form in a stepwise manner whereas the 30 S phosphofructokinase appeared to represent a stable polymer of the enzyme.

The reports of Hussey et al.[85] and Uyeda et al.[73] were conflicting. Hussey et al. found a higher phosphate content in the phosphofructokinase eluted in the trailing fractions from the Sepharose column. These fractions were expected to contain phosphofructokinase of lower molecular weight. The average molecular weight of the phosphofructokinase with high phosphorylation grade, as described by Uyeda et al., was distinctly higher than that of the low phosphate form. Obviously, the low and high phosphate forms of phosphofructokinase isolated by both research groups were not the same. It remains an open question whether or not the different forms of phosphofructokinase were originally present in the muscle material or were generated during the purification procedures. Since both research groups used different purification methods, the second possibility appears to be more probable.

Experiments with phosphofructokinase preparations from *Ascaris suum* phosphorylated to different extents[153] revealed no differences in the molecular weight distribution as determined by high pressure liquid chromatography. We have also tried to correlate differences in the aggregation behavior of various preparations of rabbit muscle phosphofructokinase to the phosphate content of the enzyme by a method developed in this laboratory.[83] However, these attempts did not provide a clear picture of events. This may be due, in part, to the complex association behavior of phosphofructokinase which is not easily described by a simple model.

B. Effects of Phosphorylation on the Kinetic Properties of Phosphofructokinase

1. Liver Phosphofructokinase

As mentioned above, the decrease of phosphofructokinase activity in livers from starved rats was one of the experimental observations that have led to the investigation of phosphorylation of liver phosphofructokinase.[26] However, the phosphate content of phosphofructokinase was not measured in these early experiments, and from the present point of view, there is evidence that the decrease in activity of liver phosphofructokinase upon starvation of the rats was caused by metabolite controlled rather than by phosphorylation controlled changes in the association state of the enzyme.

When phosphofructokinase from hepatocytes or liver[86,87] was assayed in the presence of subsaturating concentrations of fructose-6-phosphate, a decrease of activity was observed following the administration of glucagon. These changes were attributed to a shift in the sigmoid saturation curve for fructose-6-phosphate towards higher concentrations of the substrate. No effect was observed when phosphofructokinase activities were determined in the presence of saturating substrate concentrations. Since the effects of glucagon administration were not abolished by gel filtration or ammonium sulfate precipitation, it appeared convincing to attribute the changes of the kinetic properties of phosphofructokinase to an increase in phosphofructokinase phosphorylation which was also a consequence of the application of glucagon.[28]

The discovery of fructose-2,6-bisphosphate as a potent activator of phosphofructokinase which is tightly bound to the enzyme has revealed a new perspective on these experiments. The decrease of phosphofructokinase activity in the presence of subsaturating concentrations of fructose-6-phosphate, as induced by glucagon, may reflect the diminished concentration of fructose-2,6-bisphosphate (see below) rather than changes in the phosphorylation state of phosphofructokinase. In fact, Söling and Brand[38] found that the differences of fructose-6-phosphate saturation curves exhibited by phosphofructokinase from control and glucagon-treated hepatocytes were abolished by gel filtration.[38] The results obtained by these authors agree with observations in our laboratory. They likely indicated that the short-term effects on phosphofructokinase kinetics exerted by glucagon were mainly mediated by changes in the fructose-2,6-bisphosphate levels and that the phosphorylating effect on phosphofructokinase was secondary.

2. Muscle Phosphofructokinase

Following the discovery that muscle phosphofructokinase was present as a phosphoenzyme under physiological conditions,[27] the influence of phosphorylation on the kinetic properties of the enzyme was thoroughly studied. Since the in vitro kinetics of phosphofructokinase are very sensitive to changes in enzyme concentrations and to minute variations of pH, comparative kinetic studies using phosphofructokinase of different pretreatment were difficult.

Phosphofructokinase isolated from stimulated or resting muscle was different in phosphate content but showed congruent saturation curves for fructose-6-phosphate and equal sensitivities towards inhibition by ATP and citrate. Under these conditions, the sensitivity for activation by 5′-AMP and glucose-1,6-bisphosphate was also unaffected.

Foe and Kemp[61] studied the kinetic properties of purified phosphofructokinase after in vitro dephosphorylation with intestinal phosphatase and rephosphorylation to one phosphate per phosphofructokinase subunit with cAMP-dependent protein kinase. The kinetic measurements were performed in the presence of 150 m*M* KCl. As shown by Figure 9, there was a shift of the fructose-6-phosphate saturation curve to higher concentrations when the assays were performed with the phosphorylated phosphofructokinase. The dependence of the inhibitory effect of ATP on its concentration was only slightly different between phosphorylated and dephosphorylated phosphofructokinase, but corresponded to the fructose-6-phosphate saturation curve as the phosphorylated phosphofructokinase required slightly less ATP for a given extent of inhibition. The results indicated that phosphorylated phosphofructokinase was slightly stronger inhibited by ATP and was less susceptible to activation by 5′-AMP, phosphate, and fructose-2,6-bisphosphate.

The findings of Foe and Kemp[61] would suggest that dephosphorylation of phosphofructokinase favored activity under physiological conditions. However, close inspection of the reported data may raise the question whether the experiments were performed at conditions which were exactly identical. For instance, the relative activities, measured in the presence of 0.5 m*M* ATP and 2.0 m*M* fructose-6-phosphate, differed more between two series of experiments performed with the same enzyme preparation than between phosphorylated and dephosphorylated phosphofructokinase. Since variations of the pH values by 0.01 units may have exerted changes in the properties of phosphofructokinase similar to those reported for the phospho- and dephospho- forms of phosphofructokinase, the experimental difficulties involved in the comparison of the properties of phosphofructokinase preparations should be quite obvious. Unfortunately, the clear interpretation of the results of kinetic measurements is compromised by the fact.

A different approach to elucidate changes of phosphofructokinase kinetics by phosphorylation was employed by Krystek and Hofer.[74] As discussed above, trypsin specifically cleaved the C–terminal octapeptide containing the phosphorylated serine residue without affecting the remaining protein molecule and enzyme activity as measured under optimum conditions. Since the tryptic release of the phosphopeptide was inhibited by the presence of NH_4^+ and K^+ ions, an interaction of the phosphopeptide with the monovalent cations appeared possible.

Phosphofructokinase from rabbit skeletal muscle exhibited only low activity in the absence of either NH_4^+ or K^+ ions. This holds also in the presence of other positive effectors of the enzyme. Figure 10 shows the saturation curves for ammonium ions of native phosphofructokinase and of a phosphofructokinase sample from which the phosphopeptide was cleaved by trypsin. The saturation curve of the enzyme preparation devoid of the C-terminal peptide was distinctly shifted to higher concentrations of the activator. The quantitative comparison of the saturation curves obtained with untreated and trypsin incubated enzyme was compromised by their sigmoid character.

The presence of glucose-1,6-bisphosphate in a saturating concentration (0.2 m*M*) changed the saturation curves for NH_4^+ from sigmoid to hyperbolic. Figure 11 shows double reciprocal plots of the effect of various NH_4^+ concentrations on phosphofructokinase activity before and after the cleavage of the phosphorylated C-terminal peptide by trypsin. The experiments were performed at pH 6.9. The maximum velocity, which was equal at pH 7.6, was lower for the trypsin-treated phosphofructokinase by a factor of 1.16 at pH 6.9. Yet, a greater difference was observed for the values of half-saturating ammonium ion concentrations which increased from 1.6 m*M* in the case of the untreated phosphofructokinase to 10.5 m*M* when the trypsin-treated phosphofructokinase was assayed.

K^+ had a similar effect as NH_4^+ on phosphofructokinase. The decrease in the affinity to ammonium ions was paralleled by a decrease in the affinity towards K^+. The concentration of K^+ required for half-saturation was about five to ten times higher than that required for half-saturation of phosphofructokinase with ammonium ions.

The changes in the affinity towards the activating cations, induced by the release of the C-terminal phosphopeptide from phosphofructokinase, suggested that the phosphorylated peptide regulated the binding of these activators of phosphofructokinase. On the other hand, the phosphopeptide appeared not to be a part of this binding domain itself. The absence of the C-terminal peptide in phosphofructokinase did not prevent binding of monovalent cations. A mechanistic explanation of the experimental results could be based on the assumption of an inhibition of cation binding by the positive charges of the basic C-terminal peptide. The negative charges of the phosphate residue could neutralize the positive charges of arginine and lysine residues, and thus, reversible phosphorylation and dephosphorylation could provide a mechanism for the regulation of phosphofructokinase activity in muscle.

The phosphopeptide region of phosphofructokinase from rat liver was also susceptible to limited proteolysis as demonstrated by Pilkis et al.[124] Removal of the phosphopeptide did

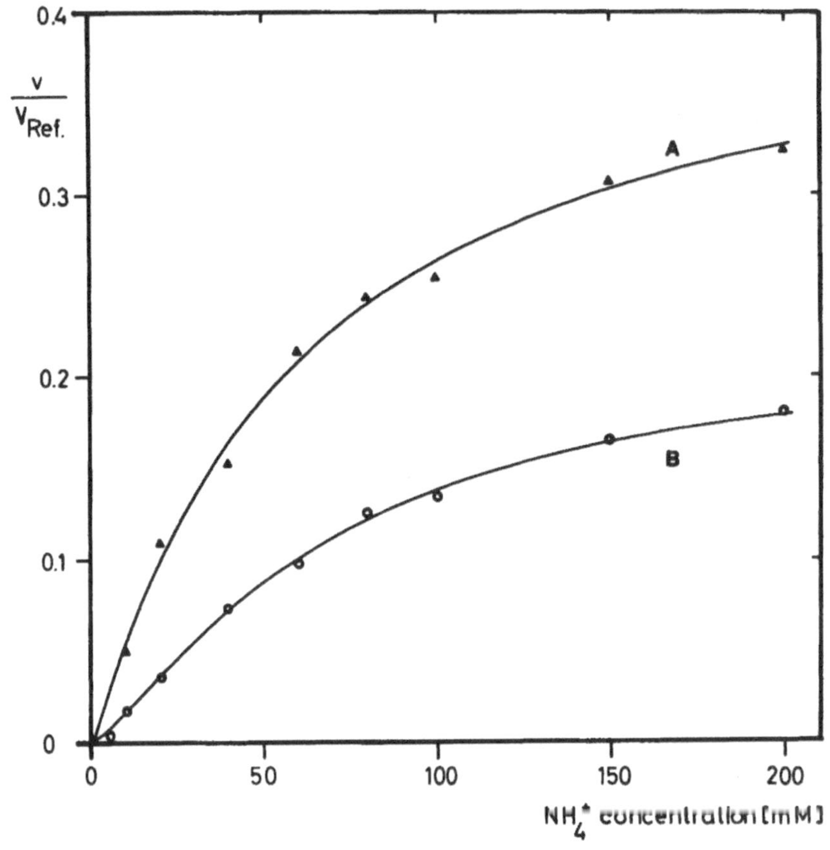

FIGURE 10. Influence of varying ammonium concentrations on the activities of native (A) and trypsin-treated (B) phosphofructokinase. The experiments were performed at pH 6.9 in the presence of 4.0 m*M* fructose-6-phosphate, 2.0 m*M* ATP, and 4.0 m*M* $MgCl_2$.

not significantly affect the molecular weight of the subunits of the enzyme nor the catalytic activity measured under optimum assay conditions. However, the affinities of the enzyme for the substrate fructose-6-phosphate and for the allosteric activator fructose-2,6-bisphosphate were increased. The changes of the affinities for fructose-6-phosphate and fructose-2,6-bisphosphate were in contrast to the fact that phosphorylation of the enzyme did not alter the kinetic affinities for these substances. In liver phosphofructokinase, similar to the muscle enzyme, the phosphopeptide region of the enzyme appears to be involved in the regulation of the affinity of the enzyme for substrates and effectors. Yet, the mechanism of the control and the functional role of the phosphate when bound to the serine residue in this segment of the enzyme remained unclear.

VII. PHOSPHORYLATION OF NONMAMMALIAN PHOSPHOFRUCTOKINASE

A. Effects of Phosphorylation on the Kinetic Properties of Phosphofructokinase from *Ascaris suum*

The only nonmammalian system in which phosphorylation of phosphofructokinase so far was studied is the muscle of the parasitic nematode *Ascaris suum*. The phosphofructokinase from this source was purified in the presence (+F) and absence (−F) of fluoride.[29,91] The +F enzyme contained 8.4 mol phosphate per tetramer and the −F phosphofructokinase

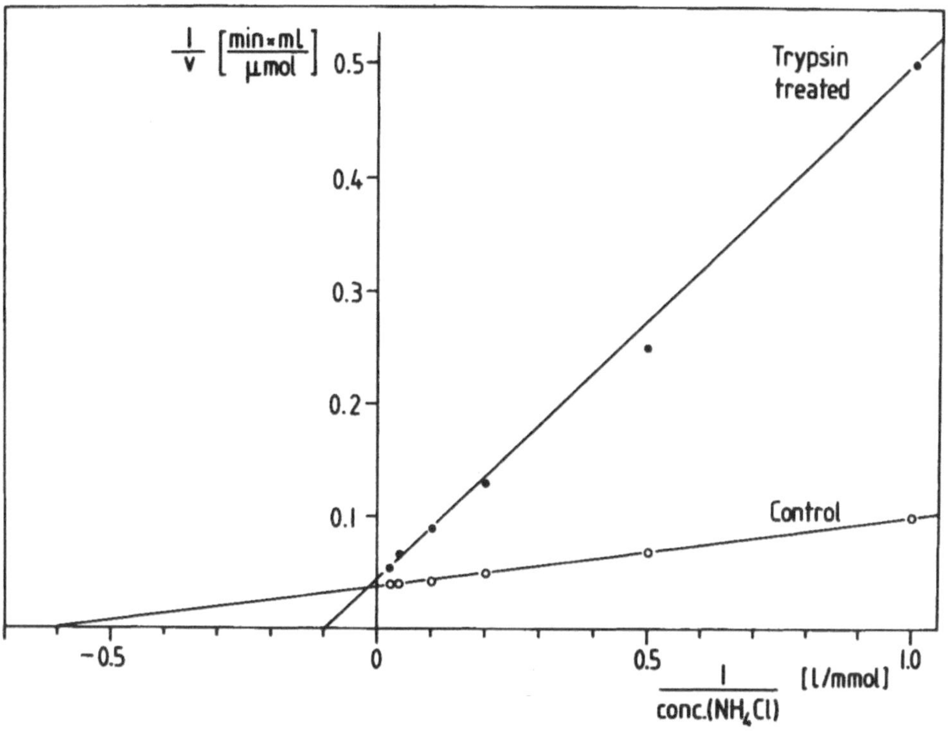

FIGURE 11. Double reciprocal plots of the saturation curves of phosphofructokinase with ammonium ions in the presence of glucose-1,6-bisphosphate. (From Krystek, E. and Hofer, H. W., *Biochem. Biophys. Res. Commun.*, 99, 1138, 1981. With permission.)

had only 3.3 mol phosphate per tetramer. Both forms of the enzyme exhibited the same specific activity when the assays were performed at optimum pH (pH 8.0) and in the presence of saturating substrate concentrations. However, the activities of both forms differed when they were assayed in the presence of concentrations of substrates and effectors comparable to those found in the muscle of the parasite (Figure 12). In the range of fructose-6-phosphate concentrations from 0.5 to 0.2 m*M*, the +F enzyme showed 1.8- to 2.5-fold higher activity than the enzyme which was purified in the absence of fluoride. These results suggested that phosphorylation of the enzyme resulted in a phosphofructokinase form which was less sensitive to inhibition by ATP, and therefore was more active in the presence of suboptimal concentrations of fructose-6-phosphate and positive effectors.

The −F enzyme was readily phosphorylated by the catalytic subunit of cAMP-dependent protein kinase from rabbit muscle or bovine heart in the presence of [γ-^{32}P]ATP. An amount of 0.02 mg of the catalytic subunit of cAMP-dependent protein kinase increased the phosphate content of 2 mg phosphofructokinase by 3 mol per mol of tetramer within 1 hr. The phosphate residues could be removed from phosphofructokinase by alkaline phosphatase from *Escherichia coli*. Over 90% of the radioactive phosphate were released within 1 hr of incubation. Analysis of the phosphate content of the dephosphorylated protein revealed 2.8 mol phosphate per mol tetramer, i.e., the phosphate residues incorporated in vitro were preferentially removed by the bacterial phosphatase. The radioactivity was released in the form of inorganic phosphate. In contrast to phosphofructokinase from rabbit muscle, the radioactive phosphate from *Ascaris* phosphofructokinase was not readily solubilized by trypsin.

Ascaris phosphofructokinase, which was phosphorylated in vitro, showed marked differences in activity from that of dephosphorylated phosphofructokinase when measured under

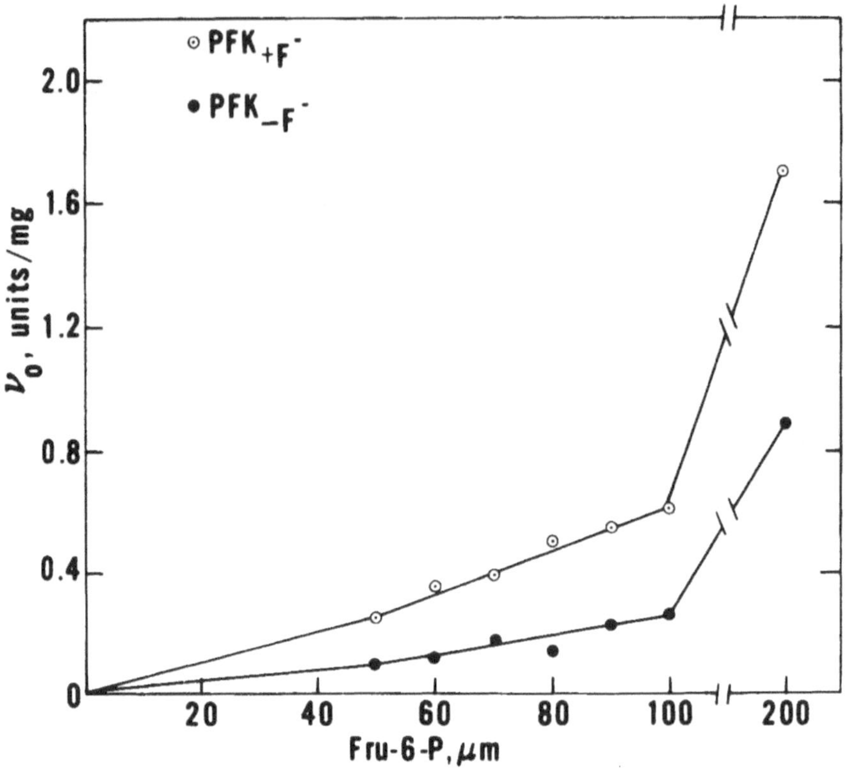

FIGURE 12. Fructose-6-phosphate saturation curves of two forms of *Ascaris* phosphofructokinase in a "Physiological Assay". The enzyme isolated in the absence of fluoride contained 2.5 times less phosphate than phosphofructokinase isolated in the presence of fluoride. (From Hofer, H. W., Allen, B. L., Kaeini, M. R., and Harris, B. G., *J. Biol. Chem.*, 257, 3807, 1982. With permission.)

regulatory conditions. Though equal amounts of phosphofructokinase exibited the same specific activity in the assays performed under optimum conditions, the activity of the phosphorylated enzyme was 250% higher than the activity of the dephosphorylated enzyme when assayed in the presence of 0.05 or 0.1 m*M* fructose-6-phosphate. Addition of the catalytic subunit of cAMP-dependent protein kinase to the assay 2 min prior to starting the reaction by the addition of fructose-6-phosphate did not significantly affect the activity of the enzyme that had previously been phosphorylated. However, the activity of the dephosphorylated enzyme was increased to the same values that were found for the phosphorylated enzyme. These results suggest that the addition of about one phosphate residue per subunit is sufficient to cause the increase in enzymatic activity of *Ascaris* phosphofructokinase.

Tryptic digestion of the radioactive labeled enzyme, followed by peptide fingerprinting and autoradiography, revealed two spots containing approximately equal radioactivity. Thus, the number of phosphate residues that can be incorporated into *Ascaris* phosphofructokinase by the action of the catalytic subunit of cAMP-dependent protein kinase in vitro as well as the pattern of the tryptic fingerprints are strong arguments for the presence of two phosphorylable sites. Since the molecular and most of the kinetic properties of the *Ascaris* phosphofructokinase are similar to mammalian phosphofructokinase, the presence of more than one phosphorylated site in the phosphofructokinase of the nematode is a subsidiary argument favoring the presence of more than one phosphorylation site in the phosphofructokinase of organisms of a higher state of evolution.

B. Properties of a Phosphofructokinase Phosphorylating Protein Kinase from *Ascaris suum*

Phosphofructokinase from *Ascaris* is effectively phosphorylated by a cAMP-dependent protein kinase which was isolated from *Ascaris* muscle[92,93] and whose catalytic subunit was partially purified.[94] The subunit has a molecular weight of 27,500; however, the native form is a dimer of double this molecular weight. The most prominent property of the protein kinase is its high affinity for *Ascaris* phosphofructokinase. The Michaelis constant for phosphofructokinase in the presence of saturating concentrations of ATP is 40 nmol/ℓ, corresponding to 0.015 mg/mℓ. This value is ten times lower than the Michaelis constant for histone which is 0.15 mg/mℓ.[94]

VIII. PHOSPHOFRUCTOKINASE AND THE REGULATION OF CARBOHYDRATE METABOLISM IN LIVER

A. Phosphofructokinase Phosphorylation and Regulatory Action of Fructose-2,6-Bisphosphate

The present view on the regulation of glycolysis in liver was completely changed by the detection of a so far unknown activator of phosphofructokinase, fructose-2,6-bisphosphate. Studies on the regulatory effects of phosphofructokinase phosphorylation have greatly stimulated the discovery of this regulatory metabolite. Since the current research on fructose-2,6-bisphosphate was most competently reviewed by Van Schaftingen and Hers[95] and Uyeda et al.,[96] this discussion will be restricted to aspects interfering with phosphofructokinase phosphorylation.

Fructose-2,6-bisphosphate is the most potent activator of phosphofructokinase that is presently known. It relieves the inhibition of the enzyme by ATP and citrate. Half maximum effects on phosphofructokinase are exerted in the range of 10^{-6} *M* concentration. Appreciable concentrations in the 10^{-5} *M* range were reported to occur in liver, brain, and heart muscle. However, the concentrations in skeletal muscle were ten times lower.

Glucagon administration to perfused rat livers or isolated hepatocytes resulted in an increase of phosphofructokinase phosphorylation. Uyeda et al.[96] have isolated two forms of phosphofructokinase containing different amounts of phosphate by chromatography on DEAE-cellulose. The low phosphate form (1.2 mol phosphate per mole tetramer) was associated with fructose-2,6-bisphosphate and was, therefore, less inhibited by ATP. The high phosphate form contained 3.3 mol phosphate per tetramer of phosphofructokinase and was free of fructose-2,6-bisphosphate when eluted from the chromatography column. When the fructose-2,6-bisphosphate was removed from the low phosphate form the difference between the low and high phosphate form, with respect to their sensitivity against ATP inhibition, disappeared. This indicates that the allosteric properties of phosphofructokinase were not affected by phosphorylation. However, the low phosphate form of the enzyme had a distinctly higher affinity for the activator fructose-2,6-bisphosphate.

B. Metabolism of Fructose-2,6-Bisphosphate

Fructose-2,6-bisphosphate was shown to be formed by the action of an enzyme called fructose-6-phosphate-2-kinase[96] or phosphofructokinase-2[95] in order to distinguish this enzyme from the phosphofructokinase acting in the glycolytic pathway. Fructose-6-phosphate-2-kinase was effectively inactivated by the administration of glucagon or epinephrine.[96,148] It was also inhibited in vitro by the action of the catalytic subunit of cAMP-dependent protein kinase in the presence of ATP. In contrast to the phosphofructokinase acting in glycolysis, phosphorylation of fructose-6-phosphate-2-kinase caused an effective down-regulation of enzyme activity. There was also evidence for the activation of fructose-2,6-bisphosphatase by phosphorylation,[96] and it was concluded that the levels of fructose-2,6-bisphosphate in tissue can be effectively regulated by phosphorylation-dephosphorylation mechanisms.[96] In combination with a decreased affinity of the phosphorylated phosphofructokinase for fructose-2,6-bisphosphate these mechanisms could provide an efficient regulation of one of the key points of carbohydrate metabolism in liver.

In gluconeogenetic organs, glycolysis and glycogenolysis are countercurrent processes in the same manner as glycolysis and gluconeogenesis from lactate or amino acids. Therefore, regulatory events must necessarily lead to opposite effects on the activity of phosphofructokinase compared to that on the activities of glycogen phosphorylase and on the key enzymes of gluconeogenesis. Activation of protein kinases, induced by alpha- or beta-receptors, activate glycogen phosphorylase and inhibit glycogen synthetase and also pyruvate kinase. Phosphorylation leads also to inactivation of fructose-6-phosphate-2-kinase and activation of fructose-2,6-bisphosphatase.[96] Both regulatory effects would result in a decrease of fructose-2,6-bisphosphate concentration. Furthermore, phosphofructokinase is phosphorylated. Though this phosphorylation does not induce obvious changes in the kinetics of the enzyme, it seems to reduce the affinity of phosphofructokinase for the activator fructose-2,6-bisphosphate and finally results in reduced glycolytic rate. Therefore, cAMP- and Ca^{2+}-dependent phosphorylation mechanisms in liver are in favor of the formation of glucose by activating glycogenolysis and gluconeogenesis, as well as by the inhibition of the glycolytic pathway.

IX. PHOSPHOFRUCTOKINASE PHOSPHORYLATION AND THE REGULATION OF CARBOHYDRATE METABOLISM IN MUSCLE

A. Metabolic Requirements for Regulation of Phosphofructokinase in Muscle

The metabolic basis for the concerted regulation of glycogenolysis and glycolysis in muscle is quite different from that in liver. In liver, glycogen serves as a store for the generation of blood glucose during interprandial periods. Therefore, glycogenolysis and glycolysis must be regulated in opposite manner. In muscle, especially in the fast-twitch glycolytic fibers, glycogen is the main fuel for the generation of phosphate bond energy, and the regulation of the glycogenolytic and glycolytic pathways should be positively coordinated. Mechanisms, as described for the regulation of the fructose-2,6-bisphosphate level and of glycolysis in liver, would in muscle lead to effects adverse to the physiological requirements. This holds especially for the inactivation of fructose-6-phosphate-2-kinase and the concomitant activation of fructose-2,6-bisphosphate phosphatase by phosphorylation. The consecutive decrease in fructose-2,6-bisphosphate concentration would reduce phosphofructokinase activity in metabolic situations which favor glycogen breakdown and hydrolysis of ATP in the course of muscle contraction. However, it was clearly demonstrated by perfusion experiments with the rat hindlimb[131] that the formation of lactate and aerobic glycolysis were stimulated under all conditions that favored glycogen degradation in muscle, e.g., muscle contraction or adrenergic stimulation.

Richter et al.[131] observed a four- to fivefold increase of glycogen degradation in fast-twitch muscles and a sevenfold increase of lactate formation by the perfused hindlimb in the presence of epinephrine as compared to resting controls. Stimulation of the muscles to contract induced a 25- to 30-fold rise of glycogen degradation, as well as a 10-fold increase of lactate formation and a 3-fold higher oxygen uptake indicating a corresponding stimulation of aerobic glycolysis. The experiments clearly demonstrated that phosphofructokinase activity in muscle was stimulated under all physiological situations in which cAMP- or Ca^{2+}-dependent protein kinases were activated.

The metabolism of fructose 2,6-bisphosphate in skeletal muscle has not been extensively studied. Uyeda et al.[96] reported a fructose-2,6-bisphosphate concentration in muscle of 2.6 $\pm$ 0.6 nmol/g. This concentration was ten times lower than the concentration of the effector in liver and appears to be in the same order of magnitude as the concentration of phosphofructokinase. The activity of the fructose-6-phosphate-2-kinase in muscle was 0.19 m units per gram. Therefore, the activity of this enzyme in muscle was also 10- to 20-fold lower than in liver. Obviously, there was no correlation between the content of fructose-2,6-bisphosphate and the glycolytic turnover rates in liver and muscle. However, it was also demonstrated that phosphofructokinase from muscle was more sensitive to activation by fructose-2,6-bisphosphate compared to the liver enzyme.[98] The $K_{0.5}$ for activation of muscle phosphofructokinase was between 2.5- and 4-fold lower than those observed for the liver enzyme. The data suggest that fructose-2,6-bisphosphate could render muscle phosphofructokinase in a partly activated state in vivo. Presently, however, no data are available providing the mechanisms of regulation of the fructose-2,6-bisphosphate levels in muscle in coordination with the functional state.

B. The Possible Role of Phosphofructokinase Phosphorylation in the Regulation of Glycolysis in Muscle

The experimental data, indicating stimulation of the glycolytic flux by muscle contraction and also by adrenergic agonists, are at variance with the finding of increased sensitivity of phosphorylated rabbit muscle phosphofructokinase against ATP inhibition.[61] Since phosphofructokinase phosphorylation is increased during muscle contraction,[64,75] it should not result in an amplification of phosphofructokinase inhibition. An increased phosphorylation of phosphofructokinase in response to stimuli causing muscle contraction, like acetylcholine, was also found in *Ascaris*.[93] In the case of the phosphofructokinase of the nematode, stimulation of enzyme activity by phosphorylation was evident.[29]

Experiments performed with a trypsin-modified phosphofructokinase, from which the phosphopeptide was released, were in favor of another theory for the influence of phosphorylation of muscle phosphofructokinase on the regulation of glycolysis.[74] As has been shown by several authors,[99-102] phosphofructokinase is activated by ammonium ions even in the presence of K^+ ions. Ammonium ions are produced during muscle contraction[103] by deamination of 5′-AMP.[104] The ATP concentrations in muscle are at a fairly constant level, even during contraction, due to the action of creatin kinase and adenylate kinase. The concerted action of both enzymes enable muscle cells to mobilize two energy-rich phosphate bonds of ATP yielding 5′-AMP. High concentrations of 5′-AMP inhibit the disproportionation of ADP to ATP and 5′-AMP which is catalyzed by adenylate kinase. Therefore, the 5′-AMP is deaminated to IMP by adenylate deaminase in a reaction sequence that is a part of the so-called Lowenstein cycle.[104] The mechanism leads to the release of considerable amounts of ammonia during muscle contraction and rises the NH_4^+ concentration more than tenfold.[149] Concomitant phosphorylation of phosphofructokinase, increasing the affinity of the enzyme for activating ammonium ions, would synergistically stimulate the glycolytic flux.

C. The Possible Interaction of Phosphofructokinase Phosphorylation and the Activation of the Enzyme by Glucose-1,6-bisphosphate

Since the role of fructose-2,6-bisphosphate in the regulation of phosphofructokinase in muscle remains unclear, another effector of phosphofructokinase retains its central importance. Glucose-1,6-bisphosphate, next to fructose-2,6-bisphosphate, is the most efficient stimulator of phosphofructokinase.[15] It can be produced by a side reaction of phosphofructokinase,[106-108] and stimulation of phosphofructokinase would also lead to increased synthesis of glucose-1,6-bisphosphate. As pointed out by Beitner,[109-120] glucose-1,6-bisphosphate is also able to regulate several other enzymes of carbohydrate metabolism in addition to its effect on phosphofructokinase. With respect to the increase of the glucose-1,6-bisphosphate levels by epinephrine, described by Beitner et al.,[111] the deinhibition of phosphoglucomutase and of phosphofructokinase appears to be important for the coordinated stimulation of glycolysis and glycogenolysis. This view was strengthened by a recent study of Beitner et al.[150] They reported that the injection of serotonin into the rat tibialis anterior muscle increased the level of cyclic GMP and led to decreased concentrations of glucose-1,6-bisphosphate. Concomitant to the decrease in glucose-1,6-bisphosphate, the regulatory activities of phosphoglucomutase and phosphofructokinase were also reduced.

Wakelam and Pette[121] found a significant increase of glucose-1,6-bisphosphate concentrations in cultivated myotubes upon addition of epinephrine. Phosphofructokinase activity was also stimulated as indicated by a decrease of the concentration ratios of fructose-6-phosphate and fructose-1,6-bisphosphate. The authors also reported the interesting observation that the phosphatase degrading glucose-1,6-bisphosphate is activated by physiological concentrations of Ca^{2+}. The reversal of the effect by trifluoroperazine suggested the involvement of calmodulin. These findings would imply different regulation of glucose-1,6-bisphosphate concentrations in muscle cells in response to the stimulation of beta- and alpha-receptors.[122]

Considerations about the possible role of glucose-1,6-bisphosphate in the regulation of carbohydrate metabolism in muscle should also take into account the interaction of glucose-1,6-bisphosphate and ammonium ions in the activation of phosphofructokinase. In the absence of activators, the saturation curve for the activation of phosphofructokinase by NH_4^+ is strongly cooperative with a $K_{0.5}$ in the range of 10^{-2} to 10^{-3} mol/ℓ. Glucose-1,6-bisphosphate shifts the curve to physiological concentrations of NH_4^+ ions and switches it to a hyperbolic characteristic. Therefore, activation of phosphofructokinase by NH_4^+ should depend on the presence of other activators and especially on the presence of the most potent activator glucose-1,6-bisphosphate.

X. CONCLUSION

During the last decade a number of enzymes and proteins were demonstrated to be modified by reversible phosphorylation in vivo and in vitro.[30,31] In many cases, the covalent modification significantly altered the functional or molecular properties of the proteins. Hence, in the animal kingdom, phosphorylation appears to be one of the most general mechanisms of regulation of protein function. The dependence of protein kinase activity on the presence of either cAMP, Ca^{2+}, or cGMP links protein phosphorylation to the control of the second messengers of hormone action.

The phosphorylation of phosphofructokinase in liver was shown to be stimulated by glucagon, suggesting the action of cAMP-dependent protein kinase. In muscle, adrenergic stimulators led to a transient increase of the phosphorylation grade of phosphofructokinase. A more rigorous rise of the phosphorylation grade was observed as a consequence of muscle contraction, suggesting the action of a Ca^{2+}-dependent protein kinase. The predominance of a Ca^{2+}-dependent protein kinase in the phosphorylation of muscle phosphofructokinase is also indicated by the low affinity of cAMP-dependent protein kinase for muscle phosphofructokinase.[59] The total phosphate content of phosphofructokinase in muscle is in the same order of magnitude as that of other phosphoproteins, e.g., glycogen phosphorylase[56] and it can be also changed within a short period of time.[64] Therefore, muscle should be expected to contain a potent interconversion system for the enzyme in addition to cAMP-dependent protein kinase. However, attempts to isolate a highly active phosphorylating enzyme for phosphofructokinase were presently not successful.

Hormone-like substances, e.g., acetylcholine and serotonine (5-hydroxytryptamine), stimulated also the phosphorylation of phosphofructokinase in *Ascaris suum*. The action of acetylcholine is mediated by Ca^{2+} ions. Serotonine elevates the cAMP concentration in the muscle of the nematode.[93,128] Therefore, a dual action of Ca^{2+}– and cAMP-dependent protein kinases in phosphorylation of phosphofructokinase must be assumed also in the case of *Ascaris* muscle.

In mammalian skeletal muscle and in the muscle of the nematode, the events which were shown to increase phosphofructokinase phosphorylation also stimulated the rate of glycolysis. This rate was equivalent to the substrate turnover in the rate-limiting step, i.e., the phosphofructokinase reaction. Thus, there was a striking coincidence between the catalytic rate of phosphofructokinase and the extent of its phosphorylation. For *Ascaris* phosphofructokinase, the transition by phosphorylation into a higher active state was also demonstrated in vitro. Therefore, this phosphofructokinase augments the list of enzymes fulfilling the criteria for interconvertible systems defined by Krebs.[30]

Effects of phosphorylation on regulatory kinetics were less clear in the case of mammalian phosphofructokinase. In liver which is predominantly a gluconeogenetic organ, phosphorylation of phosphofructokinase reduced its capability to bind the regulating metabolite fructose-2,6-bisphosphate. From a teleologic point of view, this phenomenon would justify the existence of the phosphorylation mechanism. However, reduction of phosphofructokinase activity by phosphorylation is not suggestive for muscle. Since phosphofructokinase from liver and muscle are different isoenzymes, it is not necessary to postulate analogous consequences of phosphorylation. A view into the biochemical literature teaches us that evolution has always created regulatory mechanisms of enzymes that meet the physiological demands of cells and organs. An overorganization in the sense of useless regulation may occur on the level of complex structures and may even lead to pathologic failures. Yet, excessive complexity, not justified by functional demands, was never unambiguously demonstrated on the level of an enzyme system. There are several examples that interconversion is restricted to a single isoenzyme when isoenzymes exert their catalytic function under different metabolic requirements. The phosphorylation of liver pyruvate kinase by cAMP-dependent protein kinase results in a reduction of enzyme activity under physiological conditions.[131,132] Inhibition of the formation of pyruvate from phosphoenolpyruvate is a prerequisite for effective gluconeogenesis. However, in the mainly glycolytic organ muscle pyruvate kinase is not a substrate for cAMP-dependent protein kinase and the enzyme is not phosphorylated in vivo. This example strengthens the idea that phosphorylation mechanisms have only evolved when they were useful for physiological requirements.

Ca^{2+}- and cAMP-dependent activation of muscle phosphofructokinase by phosphorylation would be a mechanism that is in concordance with metabolic requirements. At the present, the activation has not been conclusively demonstrated for the intact enzyme in vitro. On the other hand, phosphorylation of phosphofructokinase may also be significant for the regulatory function of phosphofructokinase when it does not change the regulatory kinetics in vitro. Marie and Kahn[80] have demonstrated a higher susceptibility of phosphorylated L-type pyruvate kinase to proteolytic breakdown. The half-life of phosphofructokinase in skeletal muscle under normal conditions is less than 1 day.[133] The synthesis of the enzyme, as measured by the incorporation of [1-^{14}C]leucine, was inhibited by more than 90% in the presence of cycloheximide.[64,75] However, the phosphofructokinase activity in muscle was not changed, indicating that the proteolytic degradation of phosphofructokinase was also inhibited by cycloheximide. The incorporation of radioactive phosphate into phosphofructokinase was unchanged as compared to the controls, which were not treated with cycloheximide. These experiments clearly showed that phosphofructokinase phosphorylation is a process independent of protein synthesis and protein degradation, and there is no evidence that proteolytic degradation of phosphofructokinase is affected by phosphorylation of the kinase. Moreover, the cellular concentrations of phosphofructokinase in muscle were shown to be fairly constant in relation to the concentrations of other glycolytic enzymes.[136,137] Therefore, specific regulation of the cellular phosphofructokinase concentration by activation of degradation mechanisms independent of the turnover of the other enzymes of the glycolytic pathway appears unlikely.

Another hypothetical possibility was discussed by Söling and Brand.[38] Phosphofructokinase phosphorylation could also affect the glycolytic flux rate by changing the spatial arrangement of glycolytic enzymes in muscle. Supramolecular organization of glycolytic enzymes could facilitate the transfer of the product of a reaction to the following enzyme and prevent the dilution of metabolites over the whole cellular space. Though there are some arguments in favor of supramolecular organization in the cells,[123] an influence of phosphorylation of phosphofructokinase on its tendency to interact with other macromolecules and cellular structures[125,126] was not verified so far. Since interactions of phosphofructokinase with other macromolecules were shown to change its kinetic properties,[127,128] an influence of phosphorylation on these interactions would provide a novel and elegant mechanism of metabolic regulation by enzyme interconversion.

Attempts to explain the physiological role of phosphofructokinase phosphorylation should keep in mind, however, that the failure to detect significant kinetic changes upon phosphorylation of mammalian phosphofructokinase in vitro does not rule out the existence of an effect of phosphorylation on phosphofructokinase kinetics in general. The reported kinetics of phosphorylated and nonphosphorylated phosphofructokinase were primarily studied in terms of catalytic steady-state rates in the presence of varying concentrations of substrates and effectors. These studies did not lead to a consolidated theory on the physiological *raison d' etre* of phosphofructokinase phosphorylation. There is, however, also a vast amount of literature indicating that phosphofructokinase can exist in states of different activity. Complex models were proposed to explain changes of the catalytic properties of the enzyme by structural transitions.[139-142,17] The transitions between phosphofructokinase states appear to be relatively slow.[139-141,83] Different rates of the transitions between different states of activity could, in a given physiological system and within a limited period of time, lead to considerable differences of the amount of fructose-1,6-bisphosphate generated by phosphofructokinase. Structural changes may occur with different rates in phosphorylated and dephosphorylated phosphofructokinase even when the kinetic properties of the enzyme under steady-state conditions appear similar.[144-146] Attempts to study the influence of phosphorylation on the transition rates between phosphofructokinase states with different activity have not been reported. Since it is now possible to produce phosphofructokinase with a well-defined grade of phosphorylation in vitro, the study of more complex kinetic systems with phospho- and dephospho-phosphofructokinase is indicated.

ACKNOWLEDGMENT

The author is indebted to Mrs. M. Klug and to Dr. B. G. Harris for their critical comments during the preparation of the manuscript. The studies performed in the laboratory of the author were supported by the Deutsche Forschungsgemeinschaft (Ho 650/3-5) and the Fonds der Chemischen Industrie, Frankfurt, West Germany.

REFERENCES

1. **Harden, A. and Young, W. J.,** *Proc. R. Soc. (London) Ser. B.*, 77, 405, 1906.
2. **Ostern, P., Guthke, J. A., and Terszakowec, W.,** *Hoppe-Seyler Z. Physiol. Chem.*, 243, 9, 1936.
3. **Cori, G. T. and Cori, C.,** *J. Biol. Chem.*, 116, 129, 1936.
4. **Engel'hardt, V. A. and Sakov, N. E.,** *Biokhimiya*, 8, 9, 1943.
5. **Lynen, F., Hartmann, G., Netter, K. F., and Schuegraf, A.,** *Regulation of Cellular Metabolism*, (Ciba Found. Symp.), Churchill, London, 1959, 256.
6. **Passonneau, J. V. and Lowry, O. H.,** *Biochem. Biophys. Res. Commun.*, 7, 10, 1962.
7. **Salas, M. L., Vinuela, E., Salas, M., and Sols, A.,** *Biochem. Biophys. Res. Commun.*, 19, 571, 1965.
8. **Bucher, T. and Russmann, W.,** *Angew. Chem.*, 75, 881, 1963.
9. **Ruderman, N. B., Houghton, C. R. S., and Hems, R.,** *Biochem. J.*, 158, 191, 1976.
10. **Berger, M., Hagg, S., Goodman, M. N., and Ruderman, N. B.,** *Biochem. J.*, 158, 191, 1976.
11. **Lardy, H. A. and Parks, R. E. Jr.,** *Enzymes: Units of Biological Structure and Function*, Academic Press, New York, 1956, 584.
12. **Passonneau, J. V. and Lowry, O. H.,** *Biochem. Biophys. Res. Commun.*, 13, 372, 1963.
13. **Vinuela, E., Salas, M. L., and Sols, A.,** *Biochem. Biophys. Res. Commun.*, 12, 140, 1963.
14. **Sols, A.,** *Curr. Top. Cell. Regul.*, 19, 77, 1981.
15. **Hofer, H. W. and Pette, D.,** *Hoppe-Seyler Z. Physiol. Chem.*, 49, 1105, 1968.
16. **Hulme, E. C. and Tipton, K. F.,** *FEBS Lett.*, 12, 197, 1971.
17. **Hofer, H. W.,** *Hoppe-Seyler Z. Physiol. Chem.*, 352, 997, 1971.
18. **Lad, P. M. and Hammes, G. G.,** *Biochemistry*, 13, 4530, 1974.
19. **Lad, P. M., Hill, D. E., and Hammes, G. G.,** *Biochemistry*, 12, 4303, 1973.
20. **Paradies, H. A.,** *J. Biol. Chem.*, 254, 7495, 1979.
21. **Paetkau, V. and Lardy, H. A.,** *J. Biol. Chem.*, 242, 2035, 1967.
22. **Kemp, R. G. and Forest, P. B.,** *Biochemistry*, 7, 2596, 1968.
23. **Hofer, H. W.,** *Hoppe-Seyler Z. Physiol. Chem.*, 351, 649, 1970.
24. **Mathias, M. M. and Kemp, R. G.,** *Biochemistry*, 11, 578, 1972.
25. **Gilbert, H. F.,** *J. Biol. Chem.*, 257, 12086, 1982.
26. **Brand, I. A. and Söling, H. D.,** *FEBS Lett.*, 57, 163, 1975.
27. **Hofer, H. W. and Furst, M.,** *FEBS Lett.*, 62, 118, 1976.
28. **Kagimoto, T. and Uyeda, K.,** *J. Biol. Chem.*, 254, 5584, 1979.
29. **Hofer, H. W., Allen, B. L., Kaeini, M. R., and Harris, B. G.,** *J. Biol. Chem.*, 257, 3807, 1982.
30. **Krebs, E. G. and Beavo, J. A.,** *Annu. Rev. Biochem.*, 48, 923, 1979.
31. **Chock, P. B., Rhee, S. G., and Stadtman, E. R.,** *Annu. Rev. Biochem.*, 49, 813, 1980.
32. **Mansour, T. E.,** *Curr. Top. Cell. Regul.*, 5, 2, 1972.
33. **Bloxham, D. P. and Lardy, H. A.,** *The Enzymes*, Vol. 8 A, 2nd ed., Boyer, P., Ed., Academic Press, New York, 1973, 239.
34. **Ramaiah, A.,** *Curr. Top. Cell. Regul.*, 8, 298, 1974.
35. **Hofmann, E.,** *Rev. Physiol. Biochem. Pharmacol.*, 75, 1, 1976.
36. **Uyeda, K.,** *Adv. Enzymol.*, 48, 193, 1979.
37. **Goldhammer, A. R. and Paradies, H. H.,** *Curr. Top. Cell. Regul.*, 15, 109, 1979.
38. **Söling, H. D. and Brand, I. A.,** *Curr. Top. Cell. Regul.*, 20, 107, 1981.
39. **Mansour, T. E. and Mansour, J. M.,** *J. Biol. Chem.*, 237, 629, 1962.
40. **Mansour, T. E., Sutherland, E. W., Rall, T. W., and Bueding, E.,** *J. Biol. Chem.*, 235, 466, 1960.
41. **Vinuela, E., Salas, M. L., Salas, M., and Sols, A.,** *Biochem. Biophys. Res. Commun.*, 15, 243, 1964.
42. **Atzpodien, W., Gancedo, J. M., Hagmaier, V., and Holzer, H.,** *Eur. J. Biochem.*, 12, 6, 1970.
43. **Atzpodien, W. and Bode, H.,** *Eur. J. Biochem.*, 12, 126, 1970.
44. **Afting, E. G., Ruppert, D., Hagmaier, V., and Holzer, H.,** *Arch. Biochem. Biophys.*, 143, 587, 1971.
45. **Afting, E. G., Ruppert, D., and Holzer, H.,** *Arch. Biochem. Biophys.*, 152, 433, 1972.
46. **Karadseh, N. S., Ananthanarayanan, M., and Ramaiah, A.,** *Biochem. Biophys. Res. Commun.*, 57, 771, 1974.
47. **Ramaiah, A. and Tejwani, G. W.,** *Eur. J. Biochem.*, 39, 183, 1973.
48. **Mansour, T. E.,** *J. Biol. Chem.*, 247, 6059, 1972.
49. **Clark, M. G. and Patten, G. S.,** *J. Biol. Chem.*, 256, 27, 1981.
50. **Clark, M. G. and Patten, G. S.,** *Nature (London)*, 292, 461, 1981.
51. **Clark, M. G., Filsell, O. H., and Patten, G. S.,** *J. Biol. Chem.*, 257, 271, 1982.
52. **Diamond, I., Legg, A., Schneider, J. A., and Rozengurt, E.,** *J. Biol. Chem.*, 253, 872, 1978.
53. **Schneider, J. A., Diamond, I., and Rozengurt, E.** *J. Biol. Chem.*, 253, 872, 1978.
54. **Cohen, P.,** *Curr. Top. Cell. Regul.*, 14, 117, 1978.
55. **Kirschenlohr, H. L. and Hofer, H. W.,** *Fed. Eur. Biochem. Soc. Meet., 13th*, Abstr. C2-P51-051, 1980.

56. **Kirschenlohr, H. L. and Hofer, H. W.,** *Arch. Biochem. Biophys.*, 220, 637, 1983.
57. **Uyeda, K.,** *J. Biol. Chem.*, 245, 2268, 1970.
58. **Hanson, R. L., Rudolph, F. B., and Lardy, H. A.,** *J. Biol. Chem.*, 248, 7852, 1973.
59. **Kemp, R. G., Foe, L. G., Latshaw, S. P., Poorman, R. A., and Heinrikson, R. C.,** *J. Biol. Chem.*, 256, 7282, 1981.
60. **Hofer, H. W. and Soerensen-Ziganke, B.,** *Fed. Eur. Biochem. Soc. Meet., 13th,* Abstr. C2-P52-052, 1980.
61. **Foe, L. G. and Kemp, R. G.,** *J. Biol. Chem.*, 257, 6368, 1982.
62. **Antoniw, J. F., Nimmo, H. G., Yeaman, S. J., and Cohen, P.,** *Biochem. J.*, 162, 423, 1977.
63. **Varsanyi, M. and Heilmeyer, L. M. G.,** *Biochemistry*, 18, 4869, 1979.
64. **Hofer, H. W. and Soerensen-Ziganke, B.,** *Biochem. Biophys. Res. Commun.*, 90, 199, 1979.
65. **Kagimoto, T. and Uyeda, K.,** *Arch. Biochem. Biophys.*, 203, 792, 1980.
66. **Lagrange, J. L., Marie, J., Cottreau, D., Fischer, S., and Kahn, A.,** *Biochim. Biophys. Acta*, 612, 213, 1980.
67. **Mendicino, J., Leibach, F., and Reddy, S.,** *Biochemistry*, 17, 4664, 1978.
68. **Riquelme, P. T., Fox, R. W., and Kemp, R. G.,** *Biochem. Biophys. Res. Commun.*, 81, 869, 1978.
69. **Riquelme, P. T., Hosey, M. M., Marcus, F., and Kemp, R. G.,** *Biochem. Biophys. Res. Commun.*, 85, 1480, 1978.
70. **Riquelme, P. T. and Kemp, R. G.,** *J. Biol. Chem.*, 255, 4367, 1980.
71. **Brand, I. A. and Söling, H. D.,** *Eur. J. Biochem.*, 122, 175, 1982.
72. **Soerensen-Ziganke, B. and Hofer, H. W.,** *Biochem. Biophys. Res. Commun.*, 90, 204, 1979.
73. **Uyeda, K., Miyatake, A., Luby, L. J., and Richards, E. G.,** *J. Biol. Chem.*, 253, 8319, 1978.
74. **Krystek, E. and Hofer, H. W.,** *Biochem. Biophys. Res. Commun.*, 99, 1138, 1981.
75. **Hofer, H. W.,** *Proteins: Structure, Function and Industrial Application,* Hofmann, E., Pfeil, W., and Aurich, H., Eds., Pergamon Press, Oxford, 1978, 413.
76. **Emerk, K. and Frieden, C.,** *Arch. Biochem. Biophys.*, 164, 233, 1974.
77. **Kemp, B. E., Benjamini, E., and Krebs, E. G.,** *Proc. Natl. Acad. Sci. U.S.A.*, 72, 3448, 1975.
78. **Maller, J. L., Kemp, B. E., and Krebs, E. G.,** *Proc. Natl. Acad. Sci. U.S.A.*, 75, 248, 1978.
79. **Vora, S., Seaman, C., Durham, S., and Piomelli, S.,***Proc. Natl. Acad. Sci. U.S.A.*, 77, 62, 1980.
80. **Marie, J. and Kahn, A.,** *Biochem. Biophys. Res. Commun.*, 94, 1387, 1980.
81. **Naijar, V. A. and Pullman, M. E.,** *Science*, 119, 631, 1954.
82. **Hofer, H. W. and Radda, G. K.,** *Eur. J. Biochem.*, 42, 341, 1974.
83. **Hofer, H. W. and Krystek, E.,** *FEBS Lett.*, 53, 217, 1975.
84. **Brand, I. A., Muller, M. K., Unger, C., and Soling, H. D.,** *FEBS Lett.*, 68, 271, 1976.
85. **Hussey, C. R., Liddle, P. F., Ardron, D., and Kelett, G. L.,** *Eur. J. Biochem.*, 80, 497, 1977.
86. **Castano, J. G., Nieto, A., and Feliu, J. E.,** *J. Biol. Chem.*, 254, 5576, 1979.
87. **Pilkis, S., Schmidt, J., Pilkis, J., and Claus, T. H.,** *Biochem. Biophys. Res. Commun.*, 88, 960, 1979.
88. **Van Schaftingen, E., Hue, L., and Hers, H. G.,** *Biochem. J.*, 192, 887, 1980.
89. **Pilkis, S. J., El-Maghrabi, M. R., Pilkis, J., Claus, T. H., and Cuming, D. A.,** *J. Biol. Chem.*, 256, 3171, 1981.
90. **Uyeda, K., Furuya, E., and Sherry, D. A.,** *J. Biol. Chem.*, 256, 8679, 1981.
91. **Harris, B. G. and Hofer, H. W.,** *Fed. Proc. Fed. Am. Soc. Exp. Biol.*, 40, 1605, 1981.
92. **Hofer, H. W., Daum, G., Thalhofer, H. P., and Harris, B. G.,** *Fed. Proc. Fed. Am. Soc. Exp. Biol.*, 41, 1953, 1982.
93. **Harris, B. G., Kaeini, M. R., and Hofer, H. W.,** *Mol. Biochem. Parasitol.*, Suppl. 1982, 4139, 1982.
94. **Thalhofer, H. P.,** Untersuchungen zur Phosphorylierung von Phosphofructokinase aus Ascaris suum, Thesis (Dipl. Biol.), Universitaet Konstanz, 1982.
95. **Hers, H. G. and Van Schaftingen, E.,** *Biochem. J.*, 206, 1, 1982.
96. **Uyeda, K., Furuya, E., Richards, C. S., and Yokoyama, M.,** *Mol. Cell. Biochem.*, 48, 97, 1982.
97. **Furuya, E., Yokoyama, M., and Uyeda, K.,** *Proc. Natl. Acad. Sci. U.S.A.*, 79, 325, 1982.
98. **Uyeda, K., Furuya, E., and Luby L. L.,** *J. Biol. Chem.*, 256, 8394, 1981.
99. **Abrahams, S. L. and Younathan, E. S.,** *J. Biol. Chem.*, 246, 2464, 1971.
100. **Tejwani, G. A., Ramaiah, A., and Ananthanarayanan, M.,** *Arch. Biochem. Biophys.*, 158, 195, 1973.
101. **Otto, M., Jacobasch, G., and Rapoport, S.,** *Eur. J. Biochem.*, 65, 201, 1976.
102. **Grosse, R., Eckert, K., Otto, M., Jacobasch, G., and Repke, K. R. H.,** *Eur. J. Biochem.*, 74, 509, 1977.
103. **Parnas, J. K.,** *Biochem. Z.*, 206, 16, 1929.
104. **Lowenstein, J. M.,** *Physiol. Rev.*, 52, 382, 1972.
105. **Hofer, H. W. and Pette, D.,** *Hoppe-Seyler Z. Physiol. Chem.*, 349, 1378, 1968.
106. **Hofer, H. W. and Pette, D.,** *Life Sci.*, 5, 199, 1966.
107. **Eyer, P. and Pette, D.,** *Life Sci.*, 6, 191, 1967.
108. **Eyer, P., Hofer, H. W., Krystek, E., and Pette, D.,** *Eur. J. Biochem.*, 20, 153, 1971.

109. **Beitner, R.,** *Trends Biochem. Sci.*, 4, 228, 1979.
110. **Beitner, R., Haberman, S., and Livni, L.,** *Biochim. Biophys. Acta* 397, 355, 1975.
111. **Beitner, R., Haberman, S., and Nordenberg, J.,** *Mol. Cell. Endocrinol.*, 10, 135, 1978.
112. **Beitner, R., Haberman, S., Nordenberg, J., and Cohen, T. J.,** *Biochim. Biophys. Acta*, 542, 537, 1978.
113. **Nordenberg, J., Klein, S., Beery, E., Kaplansky, M., and Beitner, R.,** *Int. J. Biochem.*, 13, 1005, 1981.
114. **Nordenberg, J., Heffetz, D., Cohen, T., and Beitner, R.,** *Int. J. Biochem.*, 13, 317, 1981.
115. **Beery, E., Klein, S., Nordenberg, J., and Beitner, R.,** *Biochem. Int.*, 1, 526, 1980.
116. **Beitner, R., Klein, S., and Nordenberg, J.,** *Int. J. Biochem.*, 14, 195, 1982.
117. **Nordenberg, J., Kaplansky, M., Beery, E., Klein, S., and Beitner, R.,** *Biochem. Pharmacol.*, 31, 1025, 1982.
118. **Beitner, R., Cohen, T. J., Nordenberg, J., and Haberman, S.,** *Biochim. Biophys. Acta*, 586, 266, 1979.
119. **Beitner, R., Nordenberg, J., Cohen, T. J., and Beery, E.,** *Int. J. Biochem.*, 11, 467, 1980.
120. **Beitner, R. and Cohen, T. J.,** *FEBS Lett.*, 115, 197, 1980.
121. **Wakelam, M. J. O., Emmerich, M., Pette, D.,** *Biochem. J.*, 208, 517, 1982.
122. **Wakelam, M. J. O. and Pette, D.,** *Biochem. J.*, 204, 765, 1982.
123. **Arnold, H. and Pette, D.,** *Eur. J. Biochem.*, 15, 360, 1970.
124. **Pilkis, S. J., El-Maghrabi, M. R., Pilkis, J., and Claus, T. H.,** *Arch. Biochem. Biophys.*, 215, 379, 1982.
125. **Margreth, A., Catani, C., and Schiaffino, S.,** *Biochem. J.*, 102, 1966, 1967.
126. **Karadseh, N. S. and Uyeda, K.,** *J. Biol. Chem.*, 252, 7418, 1977.
127. **Hofer, H. W. and Pette, D.,** *Life Sci.*, 4, 1591, 1965.
128. **Hofer, H. W. and Pette, D.,** *Life Sci.*, 4, 1597, 1965.
129. **Bazaes, S., Foe, L. G., and Kemp, R. G.,** *Arch. Biochem. Biophys.*, 218, 483, 1982.
130. **Oron, Y., Galasko, G., and Larner, J.,** *Mol. Cell. Biochem.*, 32, 181, 1980.
131. **Richter, E. A., Ruderman, N. B., Gavras, H., Belur, E. R., and Galbo, H.,** *Am. J. Physiol.*, 242, E25, 1982.
132. **Engstrom, L.,** *Recently Discovered Systems of Enzyme Regulation by Reversible Phosphorylation*, Cohen, P., Ed., Elsevier/North-Holland, Amsterdam, 1980, 11.
133. **Engstrom, L.,** *Curr. Top. Cell. Regul.*, 13, 29, 1978.
134. **Dolken, G. and Pette, D.,** *Hoppe-Seyler Z. Physiol. Chem.*, 355, 289, 1974.
135. **Illg, D. and Pette, D.,** *Eur. J. Biochem.*, 97, 267, 1979.
136. **Staudte, H. W. and Pette, D.,** *Comp. Biochem. Physiol.*, 41 B, 533, 1972.
137. **Pette, D. and Hofer, H. W.,** *Trends in Enzyme Histochemistry and Cytochemistry* (Ciba Found. Symp.), Excerpta Medica, Amsterdam, 1980, 231.
138. **Kemp, R. G., Tsai, M. J., and Colombo, G.,** *Biochem. Biophys. Res. Commun.*, 68, 942, 1976.
139. **Bock, P. E. and Frieden, C.,** *J. Biol. Chem.*, 251, 5630, 1976.
140. **Bock, P. E. and Frieden, C.,** *J. Biol. Chem.*, 251, 5637, 1976.
141. **Bock, P. E. and Frieden, C.,** *J. Biol. Chem.*, 251, 5644, 1976.
142. **Hood, K. and Hollaway, M. R.,** *FEBS Lett.*, 68, 8, 1976.
143. **Hollaway, M. R., Rodriguez, H. B., Hood, K., Simpson, I., and Nunn, J. D.,** *Proteins: Structure, Function and Industrial Application*, Hofmann, E., Pfeil, W., and Aurich, H., Eds., Pergamon Press, Oxford, 1978, 377.
144. **Kurganov, B. I., Dorozhko, A. I., Kagan, Z. S., and Yakolev, V. A.,** *J. Theor. Biol.*, 60, 247, 1976.
145. **Kurganov, B. I., Dorozhko, A. I., Kagan, Z. S., and Yakolev, V. A.,** *J. Theor. Biol.*, 60, 271, 1976.
146. **Kurganov, B. I., Dorozhko, A. I., Kagan, Z. S., and Yakolev, V. A.,** *J. Theor. Biol.*, 60, 287, 1976.
147. **Brand, I. A., Mieskes, G., and Soling, H. D.,** *FEBS Lett.*, 154, 65, 1983.
148. **El-Maghrabi, M. R., Claus, T. H., Pilkis, J., and Pilkis, S. J.,** *Proc. Natl. Acad. Sci. U.S.A.*, 79, 315, 1982.
149. **Goodman, M. N. and Lowenstein, J. M.,** *J. Biol. Chem.*, 252, 5054, 1977.
150. **Beitner, R., Frucht, H., and Kaplansky, M.,** *Int. J. Biochem.*, 15, 935, 1983.
151. **Krystek, E. and Hofer, H. W.,** unpublished.
152. **Hofer, H. W.,** unpublished.
153. **Kaeini, M. R., Harris, B. G., and Hofer, H. W.,** unpublished.

Chapter 6

REGULATION OF PYRUVATE KINASE IN NORMAL AND PATHOLOGICAL CONDITIONS

G. E. J. Staal and G. Rijksen

TABLE OF CONTENTS

I. INTRODUCTION

Pyruvate kinase (ATP: pyruvate 2-O-phosphotransferase, EC 2.7.1.40.) catalyzes the conversion of phospho-enol-pyruvate to pyruvate with regeneration of ATP. There are at least three mammalian isoenzymes of pyruvate kinase known, each of which is composed of four identical or nearly identical subunits.[1-24] According to the nomenclature proposed by Ibsen, these isozymes are designated as L (liver), M (muscle), and K (kidney).[1] M- and K-isozymes are also frequently designated as M_1 and M_2, respectively.[3,13] The subunit constitution of the isozymes can be designated by a subscript, e.g., M_4 for the tetrameric form of the M-isozyme and K_2M_2 to designate a hybrid with two subunits of the K-type protomer and two of the M-type protomer. The presence of hybrid isozymes has been demonstrated in many mammalian tissues, especially from rat organs.[6] Data has been reported suggesting the presence of KL hybrids in kidney and intestinal extracts and of KM hybrids in several tissues obtained from various adult mammalian species.[7,8,14,18]

The dominant isoenzyme in fetal tissue is type K.[5-9] During development the other two appear. In adult organisms type M is predominant in striated muscle and nervous tissue,[11-14] while type L predominates in liver and erythrocytes.[24] Type K is predominant in adult kidney and leukocytes.[22,23] Table 1 shows the tissue localization of the various isozymes. This table is mainly based upon studies of Imamura et al.,[15] van Berkel,[3] and Strandholm et al.[7]

The three isozymes have about similar molecular weights (200,000 to 250,000), have equal affinities for ADP, and show allosteric behavior towards phospho-enol-pyruvate.[13] The M- and K-type isoenzymes are immunologically related to each other, but there is no immunologic relationship between the M- and K-type isozymes on one hand and L-type isoenzyme on the other hand.[23] The main difference between these isoenzymes is the different regulatory behavior.[1,3] In the following sections pyruvate kinase is discussed with respect to the genetic system, the kinetic properties, especially allosteric regulation, the isozyme distribution in neoplasia and the deficiency of the L-type in patients with nonspherocytic hemolytic anemia.

II. GENETIC SYSTEM OF PYRUVATE KINASE

A. L-Type Pyruvate Kinase

Comparing the L-type found in liver with the one found in erythrocytes, differences are found with respect to molecular weight,[25] kinetics,[26,27] ability to be phosphorylated,[28-30] and electrophoresis (see Table 2).[24] However, also important similarities are observed. In the first place an antiserum raised in rabbits against pure human L-type pyruvate kinase does not discriminate between the liver and the erythrocyte enzyme.[13-24] This indicates a large structural similarity between the enzymes from these two sources. Secondly, a deficiency of the human erythrocyte pyruvate kinase is also expressed in the liver.[31-35] This is an argument for the statement that the enzymes from liver and erythrocytes are derived from one and the same gene. In order to explain the observed differences between the enzyme from liver and erythrocytes the "posttranslational modification hypothesis" was put foreward.[25] This hypothesis states that the L-type pyruvate kinase subunit is initially synthesized as an L'-subunit, with a molecular weight of 63,000. In the liver a rapid proteolysis yields the L-subunit, with a molecular weight of 60,000 resulting in L_4 pyruvate kinase. In erythrocytes proteolysis would be slow, resulting in L'_4 and $L_2'L_2$ (see Figure 1). The hypothesis was supported by the fact that the "aging" of the L'_4 and $L_2'L_2$ forms could be performed "in vitro", by a mild tryptic treatment of these enzyme forms. Such "artificial aging" resulted indeed in L_4 enzyme with the same properties as the L_4 enzyme isolated from liver. If the "posttranslational hypothesis" is correct then both in liver and reticulocytes

Table 1
THE OCCURRENCE OF THE ISOZYMES IN THE VARIOUS MAMMALIAN TISSUES

Type of isozyme	L	M	K
	Liver	Brain	Leukocytes
	Erythrocytes	Heart	Kidney
	Kidney	Muscle	Intestine
			Lung
			Fibroblasts
			Testis
			Adipose tissue
			Stomach
			Spleen

Table 2
DIFFERENCES BETWEEN HUMAN LIVER AND ERYTHROCYTE PYRUVATE KINASE

	Liver	Erythrocyte
Mol wt	60,000 (×4)	63,000 (×4)
Kinetics	"Normal"	Hysteretic response towards the substrates
$S_{0.5}$PEP	0.29 m*M*	0.62 m*M*
Phosphorylation	Easy	Difficult
Electrophoresis	1 "Fast" band	2 Slower bands

a mRNA is to be found, coding for a 63,000 mol wt L′ subunit. In order to check this, mRNA fractions from liver and erythrocytes has been isolated and translated in a cell-free reticulocyte system, in the presence of anti-proteolytic substances.[36]

The results contradicted the posttranslational modification hypothesis: two different mRNAs encode for the L′ and L pyruvate kinase. The two types of mRNAs are transcribed from a single gene, presumably involving gene rearrangement or differential processing of a common nuclear RNA precursor.[36,37] A final conclusion, however, can be reached only by determination of the complete amino acid sequences or the structures of the mRNAs and the genes of the L and L′ forms.

B. M- and K-Type Pyruvate Kinase

M-type pyruvate kinase is the major isozyme in skeletal muscle, heart, and brain;[14] the K-type isozyme predominates in tumor and fetal tissues.[5,14] Both M- and K-isozymes consist of four identical subunits. The K-subunit has been reported to be 1000 to 2000 daltons larger than the M-subunit and the two isozymes have different isoelectric points,[39] but the pI value of the K-type is shifted to that of the M-type by treatment with muscle extract.[22] Ibsen et al.[38] reported that trypsin treatment of the K-type produced a subunit with the same molecular weight as that of the M-subunit. The question of whether the M- and K-type isozymes of pyruvate kinase are synthesized by a common mRNA has been examined by isolating total RNA from rat skeletal muscle and AH-130 Yoshida ascites hepatoma cells, which express the M- and K-type isozymes, respectively, and translating the preparations in a rabbit reticulocyte lysate system.[40] The newly synthesized pyruvate kinase subunits were compared by sodium dodecylsulfate-polyacrylamide gel electrophoresis and isoelectric focusing. The pyruvate kinase subunit synthesized from hepatoma cell RNA had a slightly larger molecular

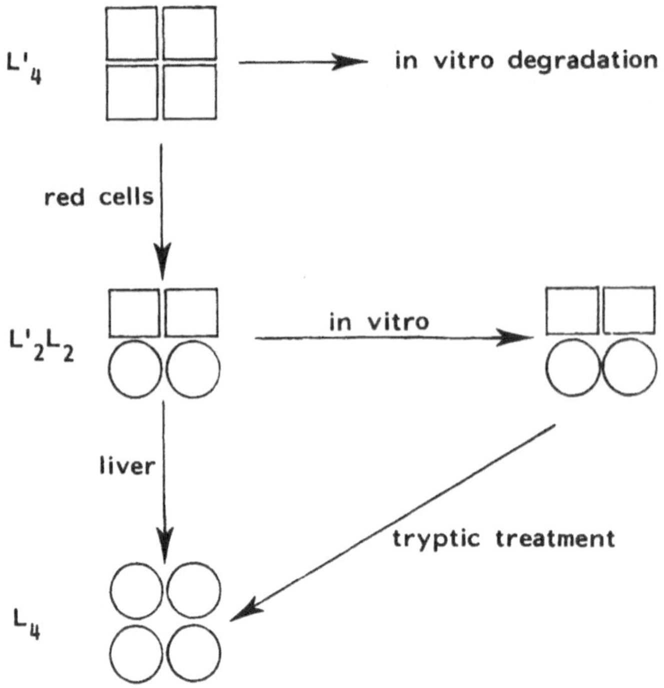

FIGURE 1. The postsynthetic evolution of L'_4 enzyme, in vivo and in vitro. □, subunit L'; ○, subunit L. (From Kahn, A., Marie, J., Garreau, H., and Sprengers, E. D., *Biochim. Biophys. Acta*, 523, 59, 1978. With permission.)

weight and higher pI value than the subunit from muscle RNA. When a mixture of the two RNAs was added to the lysate, the same result was obtained. These data indicate that the M- and K-type isozymes are translated from different messenger RNAs. It seems likely that the mRNAs for the two isozymes are transcribed from separate, but closely related, genes or the same gene. In the latter case, differential processing of RNA precursor must be responsible for the different mature mRNAs.

Further support for the hypothesis that one gene but two mRNAs are involved in the synthesis of M- and K-type isozymes came from Saheki et al.[37,41] They investigated the development of the M- and K-isozyme system during evolution. They made peptide maps of the M- and K-enzymes isolated from rabbit, rat and mouse.

M- and K-isozymes from rat have 64 peptides in common, 8 K-specific peptides, and 6 M-specific peptides. M-specific peptides are highly variable in evolution; K-specific peptides are highly stable in evolution. If one of the common peptides changes during the evolution, it remains a common peptide. In other words in such a case the change in the K-type isoenzyme is exactly the same as the change in the M-type isoenzyme. Although the latter observation is not a direct proof, it is a strong argument for the existence of only one gene for the production of M- and K-type enzymes. The considerable homology between the isozymes suggests that the two mRNAs involved in the synthesis of M- and K-subunits as reported by Noguchi, may be derived from a single gene by different processings from separate, closely related genes.[40]

Recently, the group of Kahn presented evidence that the M- and K-type pyruvate kinases are synthesized under the direction of two different RNAs.[42] Using a cell free system, they found that muscle RNA directed the translation of a pyruvate kinase subunit with a molecular weight of 61,000 daltons. The molecular weight of the M-type pyruvate kinase monomer proved to be exactly the same indicating therefore that both cell free synthesized and tissue M subunits are identical. Spleen, kidney, and lung RNA directed the cell free synthesis of a pyruvate kinase subunit with a molecular weight of 62,000 daltons that corresponds exactly to the molecular weight of the K pyruvate kinase monomer.[42]

III. KINETIC PROPERTIES OF PYRUVATE KINASE

A. Allosteric Activation and Inhibition

1. The L-Type

The central role of the liver in glucose homeostasis and the apparent involvement of L-type pyruvate kinase regulation during changes in carbohydrate flux through the pathways of glycolysis and gluconeogenesis has generated a great deal of interest in liver pyruvate kinase. The liver contains two isozymes of pyruvate kinase. The major form is the L-type isozyme and the minor form present is the K-type. The L-type is found in the parenchym-cell, the K-type in the Kupffer cell.[3] Tanaka et al.[13] have shown that the L-type is under allosteric control. The enzyme exhibits cooperativity with respect to phospho-enol-pyruvate and is activated by fructose-1,6-diphosphate which shifts the kinetics from sigmoidal to hyperbolic.[42-46] The enzymatic activity of the L-type can be modulated to a great extent.[46] The enzyme is allosterically inhibited by ATP and alanine, the inhibitory effects of which can be counteracted by the addition of fructose-1,6-diphosphate. P_i, glucose-1,6-diphosphate, and other phosphorylated hexoses influence enzyme activity too.[13] These compounds increase the apparent affinity for the substrate phospho-enol-pyruvate.[46-50]

The affinity of L-type pyruvate kinase for P_i and phosphorylated hexoses is increased by raising the phospho-enol-pyruvate concentration.[48] Lowering of the phospho-enol-pyruvate concentration results in lower affinities for P_i and phosphorylated hexoses, which are further decreased, when the inhibitors MgATP and/or alanine are also present. By comparison of the apparent affinities for various phosphorylated hexoses in the presence of ATP and alanine, Van Berkel et al.[49] were able to show that the latter two compounds introduce a different conformation probably because each binds to a different site. This conclusion was further confirmed by stability experiments at 53°C. Alanine and ATP both stabilize the enzyme at this temperature. However, when under saturating alanine concentrations ATP is added, a further stabilization occurs. These results indicate that sequential conformation changes are involved in the allosteric transition of the enzyme.

Recently, it was found that fructose-2,6-biphosphate is a potent regulator of glycolysis.[51,52] Fructose-2,6-biphosphate is a potent allosteric activator of 6-phosphofructo-1-kinase and an inhibitor of fructose-1,6-biphosphatase.[51] Since 6-phosphofructo-1-kinase is one of the most important control elements in glycolysis, it is likely that fructose-2,6-biphosphate is a significant physiological regulator of this process in liver. The activating effect of fructose-2,6-biphosphate will result in an increase of fructose-1,6-biphosphate which in turn activates pyruvate kinase. In this way fructose-2,6-biphosphate influences indirectly the activity of pyruvate kinase.

As already mentioned, pyruvate kinase of erythrocytes resembles the L-type from liver. The activity of the enzyme in red cells is also stimulated by P_i and phosphorylated hexoses, just as the L-type from liver.[53] An important difference between these two enzymes is that the hepatic enzyme is under hormonal and dietary control, whereas the erythrocyte enzyme is not.[13]

As pointed out by Sprengers et al.,[26] the L'_4 enzyme from human erythrocytes shows strong hysteretic behavior: the initial rate of enzyme activity when the enzyme is preincubated with an unsaturating concentration of phospho-enol-pyruvate is much higher than of the enzyme preincubated with ADP, at the same phospho-enol-pyruvate concentration, although the "final activity" is the same in both cases. This phenomenon was observed both in the presence and absence of fructose-1,6-diphosphate. L_4 pyruvate kinase (either from human liver or trypsin-treated L'_4 enzyme) does not show such hysteretic behavior.

2. The K- and M-Type

By comparing the inhibition of the pyruvate kinase activities of the K-type of leukocytes and the K-type of liver by amino acids it was concluded that these K-types were similar.[55] Jimenez de Asua et al.[56] showed that the K-type from rat liver is inhibited by a series of amino acids while the M-type from muscle is only inhibited by phenylalanine.[54] Pyruvate kinase of leukocytes shows cooperative interaction between the binding sites for phospho-enol-pyruvate.[55] Activity curves obtained in the presence of the allosteric inhibitor alanine are sigmoidal and are very similar to the normal activity curves obtained with the L-types of liver or erythrocytes.[44,45,53] Alanine and phenylalanine act as allosteric inhibitors. When 5 to 10 m*M* phenylalanine is present, fructose-1,6-diphosphate is not able to overcome the inhibition indicating that phenylalanine and fructose-1,6-diphosphate act as allosteric antagonists. These properties are in agreement with the two-state R $\leftrightarrows$ T model of Monod et al.[57] Normally, the leukocyte enzyme is in the R state, while alanine and phenylalanine are able to convert this R form into the T form. The effect of fructose-1,6-diphosphate can be explained by a simple reconversion of the T state to the R state.[55] It is likely that this model is also valid for the liver K-type, whereas the allosteric inhibition by phenylalanine on the activity of muscle M-type pyruvate kinase, indicates that the model may even be extended to the muscle M-type pyruvate kinase.[54]

The allosteric properties of the K-type of liver are markedly influenced by the pH.[58] At a pH lower than 7.5 hyperbolic curves are obtained; fructose-1,6-diphosphate does not affect the activity. An increase in the pH to 8.0 introduces cooperative interactions and the $K_{0.5}$ value for phospho-enol-pyruvate increases markedly. Above pH 8.0 a further increase in pH has no further affect on the allosteric properties. The presence of 1 m*M* alanine can raise the $K_{0.5}$ value for phospho-enol-pyruvate and fructose-1,6-diphosphate can transform the obtained curve into a hyperbolic relationship.

As mentioned before comparison of the K-types of liver and leukocytes leads to the conclusion that these K-types are very similar.[58] However, some differences exist. Higher concentrations of alanine are needed to inhibit the leukocyte enzyme to the same extent as the K-type from rat liver and the $K_{0.5}$ value for phospho-enol-pyruvate in the presence of 1 m*M* alanine at pH 7.5 is 0.2 m*M* for the leukocyte enzyme, whereas for the K-type from liver a value of 0.68 m*M* is obtained. These differences are probably due to a different allosteric constant (L), that is a difference in the equilibrium between the R and T state.[58]

Pogson has shown with the pyruvate kinase from adipose tissue that the presence of EDTA in the extraction buffer modifies the enzyme, resulting in an enzyme with a completely different kinetic behavior (called by Pogson Pyk-A).[59] The B form of the enzyme, isolated in the absence of EDTA, shows a high affinity for phospho-enol-pyruvate (K_m for phospho-enol-pyruvate 0.06 m*M*) whereas with the A form a $K_{0.5}$ of 0.6 m*M* has been obtained. Walker and Potter have shown that rat liver cells, cultured in the absence of glucose, contain the A form whereas in the presence of glucose the B form was obtained.[60] Van Berkel showed that the relative contribution of the two forms to the overall pyruvate kinase activity is regulated by the free Mg^{2+} concentration.[61] The B form of K-type pyruvate kinase which is present at high Mg^{2+} concentrations shows a high affinity for phospho-enol-pyruvate and its properties can be explained on the basis of the R $\leftrightarrows$ T model of Monod et al.[57] In this model fructose-1,6-diphosphate and phospho-enol-pyruvate should favor the R state whereas alanine and other inhibitory amino acids possess a higher affinity for the T state. Lowering of the Mg^{2+} concentration introduces the A form. The A form of the enzyme is characterized by a lower affinity for the substrate phospho-enol-pyruvate while this form is not sensitive to fructose-1,6-diphosphate. However, fructose-1,6-diphosphate can convert the A form into the B form which is potentiated by high Mg^{2+} concentrations.[61] In rat small intestine a considerable amount of the enzyme was found to be in the A form.[62] This observation was supported by Osterman and Fritz by electrophoretic findings of pyruvate kinase from small intestine.[63]

B. Phosphorylation

Regulation of liver L-type pyruvate kinase by a phosphoregulation mechanism has been intensively studied. Ljungström et al. were the first to report that rat liver L-type pyruvate kinase can be phosphorylated by an adenosine-3',5'-phosphate-dependent (cyclic-AMP-dependent) protein kinase.[64] The effect of cyclic-AMP-dependent phosphorylation can be reversed by a phosphoprotein phosphatase.[65] The phosphorylation was also demonstrated for the L-type from pig liver and human erythrocytes.[66,67] The phosphorylation of L-type pyruvate kinase causes important changes in the kinetics of the enzyme: decreased affinity for phospho-enol-pyruvate and increased inhibition by the allosteric inhibitors ATP and alanine.[68-70] These kinetic modifications are expected to play an important physiological role in the dynamic balance between glycolysis and gluconeogenesis. Similar changes in the kinetic properties of the enzyme are observed when isolated liver preparations are treated with glucagon, cyclic AMP, or catecholamines.[71-79] Insuline has been demonstrated to counteract the effects of glucagon and catecholamines on pyruvate kinase activity in hepatocytes and in perfused liver.[73,76] Although insulin lowers cyclic AMP levels, it is uncertain whether the activation of pyruvate kinase by insulin is solely related to changes in this nucleotide.[69]

Van Berkel et al. studied the changes in kinetic properties of L-type pyruvate kinase induced by glucagon incubation of hepatocytes or by starvation of the intact animal. The apparent affinities for the substrate phospho-enol-pyruvate and for the allosteric activator fructose-1,6-diphosphate are lowered, whereas the inhibitory actions of alanine and especially ATP are increased. These effects indicate that glucagon or starvation convert the enzyme to a form which leads to a lower pyruvate kinase activity in the isolated hepatocytes or in the liver of the intact animal. Such changes in kinetic properties were also observed by Kohl and Cottam.[82] Furthermore the change in kinetic parameters introduced by glucagon incubation of hepatocytes and starvation of the intact animal are similar and identical to the changes observed on in vitro phosphorylation of the enzyme purified from rat liver.[68-70] It might be questioned in how far the relatively small change in kinetic properties of pyruvate kinase by enzyme phosphorylation contributes to the regulation of pyruvate kinase in vivo. In this respect it is important to realize that hormones can regulate the fructose-1,6-diphosphate concentration by influencing the balance of fructose-1,6-bisphosphatase to phosphofructokinase activities. The same is valid for the concentration of glucose-1,6-diphosphate.[50] Both phosphorylated hexoses are important regulators of pyruvate kinase.[46,48,50] Therefore, hormonal control of pyruvate kinase activity is not only exerted by changes in the degree of phosphorylation of the enzyme itself but also in a more remote sense, in that hormones influence the concentration of fructose-1,6-diphosphate and glucose-1,6-diphosphate.

Blair et al. were able to demonstrate heterogeneity in the subunit composition of L-type pyruvate kinase from rat liver.[83] Isoelectric focusing revealed the presence of five bands, three of which were not influenced by phosphorylation. Despite the heterogeneity in the subunits with respect to phosphorylation the purified enzyme was capable of binding 3.7 molecules of fructose-1,6-diphosphate per tetramer. These results suggest that some of the subunits of hepatic pyruvate kinase are modified such that they lose the capacity to be phosphorylated which suggests that nonphosphorylatable forms of the enzyme may occur in the intact cell.[83]

It has been suggested that phosphorylation served as a signal for proteolytic degradation of the enzyme and that removal of the phosphorylated site was not associated with loss of the allosteric sites for ATP and fructose-1,6-diphosphate.[84] Red cell pyruvate kinase undergoes a series of proteolytic modifications during maturation of the reticulocyte. These proteolytic modifications alter the capacity of the enzyme for phosphorylation by a cAMP-dependent protein kinase.[85] Whether control of red cell pyruvate kinase can be related to that for the liver enzyme, however, is not clear at this time. The importance of proteolytic modification of liver pyruvate kinase to the physiological regulation of liver carbohydrate metabolism remains to be established.[83]

IV. ISOZYME DISTRIBUTION IN NEOPLASIA

There is increasing recognition that gene expression in tumor cells is different from that in normal cells from which the tumors originate.[86,87] In general, neoplasia is characterized by ''reprogramming of protein synthesis''.[88] After the discovery by Markert in 1959 that enzymes can exist in multimolecular forms, it became more and more evident that neoplasia is associated with changes in isozyme distribution.[89,90] The similarity of some isozyme patterns to those of fetal organs is probably one of the most remarkable features of malignant tumors.[91,92] Since isozymes differ in their primary structure isozyme alterations offer an approach to studies of abnormal genetic regulation. Fetal isozyme expression is a wide-ranging general phenomenon. Ibsen and Fishman collected 22 instances of fetal isozyme expression in tumors of various types.[93] For those proteins that appear in cancer and are also found in the fetus and various embryonal-derived tissues such as the trophoblast and the placenta, Fishman has proposed the term oncodevelopmental proteins.[93]

Although fetal isozyme expression in tumors is well established, one has to realize that many isozymes (e.g., of adult liver) undergo substantial changes in activity in response to normal dietary and hormonal alterations. So it is important to establish that the observed enzyme alterations in tumors are not consequences of metabolic changes that may occur in tumor-bearing animals. Furthermore in many cases it is still the question whether isozyme alterations are due to true changes in isozyme composition of a homogeneous cell population or to differences in the cell populations comprizing the tumor. Another question concerns the extent to which the activity levels observed are due to variations in rates of synthesis and degradation.

A. Hepatomas

The liver and its neoplastic counterpart, the hepatoma, are especially advantageous model systems for comparative studies of isozyme composition. The Morris hepatomas, chemically induced, transplantable rat neoplasms, exhibit a wide range of growth rates, degrees of differentiation, and retention of liver functions. They therefore provide an advantageous system for studies on isozyme composition as a basis of neoplastic behavior.

Tanaka et al.[94] first provided evidence for the presence of two types of pyruvate kinase in rat liver; they showed that in ascitic hepatoma AH 130 most of the enzyme is of M-type, which they believed to be present in minor amounts in normal liver too. Taylor et al.[95] showed that pyruvate kinase of the fast-growing hepatoma 3924-A was different, not only from the L-type but also from the M-type. The hepatoma enzyme was strongly inhibited by alanine, this inhibition was competitive with phospho-enol-pyruvate, in agreement with the properties of the K-type. Schapira and Gregori were able to show that the ''new isozyme'' had the same properties as the isozymes of rat liver.[96] They concluded that this ''new'' isozyme was in fact a resurgent fetal isozyme. Farina et al.[8] and Weinhouse et al.[97-99] using the Morris transplantable hepatomas and Tanaka et al.[100] and Yanagi et al.,[101] using hepatomas newly induced by chemical carcinogens, studied the pyruvate kinase isozyme patterns in neoplasms with graded degrees of differentiation. It was found that the highly differentiated tumors had isozyme patterns not necessarily different from those of normal liver. Well-differentiated tumors showed a marked decrease in the activity of the L-isozymic specific activity. The poorly differentiated tumors had little L-isozyme and showed a manyfold increase in K-isozyme specific activity.

The pyruvate kinase isozyme system appears to be a good model system to study mechanisms controlling genetic expression as was discussed by Ibsen.[1] There is a shift from the L-type toward the K-isozyme in hepatomas; the same switch occurs in regenerating liver,[102,104] in cultured hepatocytes,[103] in the liver of tumor-bearing animals,[105] in the liver of the parabiotic twin of a tumor-bearing animal,[106] in livers perfused with blood from a tumor-bearing animal,[107] in the livers of animals given injection of the blood of tumor-bearing animals,[108] and in the liver of animals given injections of extracts prepared from tumor cells.[109] It seems probably that the increase in K-isozyme activity is due to increased protein content since it can be measured immunologically.[100] The data are suggestive for the consequences of dietary and hormonal changes occurring in tumor-bearing animals. Alternatively, these observations may also suggest that a tumor-derived factor operates. Tanaka et al. suggested that this factor may be a nuclear protein factor with derepressor-like activity which can be isolated from the chromatin of tumor cells.[100] Ibsen concluded that "data concerning the K-isozyme stimulating factor are sufficient to prove that a specific effect is occurring and suggests that the factor may be a nuclear protein that functions as a derepressor."[1]

B. Brain Tumors

It has been reported that type M_4 is predominant in adult brain whereas type K_4 is present in minor amounts.[110] In the fetal brain 5 bands are observed (K_4, K_3M, K_2M_2, KM_3 and M_4), indicating that both K and M subunits are synthesized simultaneously.[110] Tolle et al.[6] reported K-M hybrids in fetal rat brain, as well as the fact that hybrids did not form upon mixing of M_4 and K_4 and electrophoresis. In the newborn both type M_4 and K_4 are present, but M_4 is predominant, whereas in the adult type M_4 is mainly present with little K_4.[110] It seems likely that during development the synthesis of M-subunits is favored.

Meningiomas and malignant gliomas are characterized by the absence of type M_4, while type K_4 and the hybrid K_3M are predominant.[110] A switch toward the K-type in human brain tumors was also observed by Bennett et al.[111] Both kinds of tumors show the same pyruvate kinase pattern, although meningiomas are slowly growing and malignant gliomas are rapidly growing tumor types.[110] This finding is in contrast with the results of Tolle et al.[6] on neuroblastomas and glioblastomas. They concluded that "tumors of neuronal origin like those from other tissues appear to revert their isozyme content and in general the greater the growth rate of the tumor, the greater the ratio of type K to type M is found." However, it is quite well possible that the cells from which meningiomas originate are characterized by the presence of the K-type. Therefore, it is not surprising that the ratio of type K to type M is not proportional to growth rate in case of meningiomas.

The K-isozyme is strongly inhibited by the amino acid L-α-alanine; this inhibition is competitive with respect to the substrate phospho-enol-pyruvate. On the contrary the M-isozyme shows no sensitivity to alanine. The correlation between alanine inhibition and isozyme composition was demonstrated by Van Veelen et al.[112] The relative intensities of the pyruvate kinase activity bands in the electropherogram can be quantitated by scanning their optical densities, and hence the proportions of K and M subunits can be calculated.[112] With the alanine inhibition test it is possible to determine the proportion of K and M subunits in normal brain and brain tumors. It should be noted that the alanine inhibition test can be performed within 10 min after a biopsy is taken and fits well within the scope of a surgical procedure.[113] Well-differentiated gliomas, i.e., astrocytomas, oligodendrogliomas, and ependymomas showed a relatively high residual pyruvate kinase activity in the presence of alanine.[114] The reverse was found for poorly differentiated gliomas. Alanine inhibition of pyruvate kinase correlates well with degree of differentiation.[114] There is also a correlation between alanine inhibition of pyruvate kinase and "one year survival" after total or subtotal resection of gliomas in adults.[115] All poorly differentiated gliomas are characterized by the presence of type K and the hybrid K_3M. In well-differentiated gliomas besides K_4 and K_3M also M_4 was present.

With some prudence one can state that in gliomas the synthesis of fetal-type pyruvate kinase is favored whereas the synthesis of the M-type (predominant in the tissue of origin) is depressed. It is not certain, however, that type K in gliomas is the same as type K in fetal brain, although they have some properties in common.

The specific activity (expressed as units per milligram protein) of pyruvate kinase in human brain increases during development.[110] In fetal brain of 12 weeks gestation, the activity was found to be 1.8 U/mg protein. At 16 weeks of age the specific activity increased to 2.1, immediately after birth to 2.8, and at adult age the specific activity was found to be 7.0 ± 0.8 U/mg protein.[110] In human brain tumors the specific activity was found to be diminished as compared to normal brain.[110] This is in contrast with pyruvate kinase activity in other tumors, e.g., hepatomas in which the enzyme activity is increased.

As already mentioned, type K pyruvate kinase is strongly inhibited by alanine. It has been reported that addition of alanine to intact Ehrlich ascites tumor cells increased the respiratory rate and decrease the glycolytic rate.[116] Concomitant with these changes is an accumulation of intracellular phospho-enol-pyruvate and ADP, and a decrease in pyruvate and ATP, indicating that pyruvate kinase is inhibited. These results are in accordance with a role of pyruvate kinase as a determinant of glycolytic activity by competing with mitochondrial oxidative phosphorylation for the available ADP.[116] Using cells, partially permeabilized to nucleotides and phosphorylated substrates, Gosalvez et al.[117] provided evidence supporting the existence in hepatocytes of a partial control by adenosine triphosphate at phosphofructokinase, which is followed by the total control by adenosine triphosphate at pyruvate kinase. The partial absence or nonoperation of this second site in Ehrlich ascites tumor cells appears to be the cause for the characteristic aerobic glycolysis, Crabtree effect, and low Pasteur effect of these cells.[117] These properties would also be of potential physiological significance in fetal tissues, which also use the K-isozyme.

C. Other Cancerous Tissues

Pyruvate kinase type K may be found in tumors of various organs. Farron et al.[118] and Nagao et al.[119] have characterized this K form in rat rhabdomyosarcoma, in which it replaces type M, the only form present in normal adult muscle. Kraaijenhagen et al.[120] studied the isozyme distribution of pyruvate kinase in lymphocytes from patients with chronic lymphocytic leukemia (CLL). The CLL lymphocytes were different from normal non-T lymphocytes with respect to pyruvate kinase residual activity in the presence of alanine. The same authors have also studied pyruvate kinase in two T cell lines and two B cell lines, compared with the enzyme of normal T and non-T lymphocytes from peripheral blood.[121] All cell types studied contained K_4 type pyruvate kinase and the hybrid K_3M. In addition, in the two T cell lines and the T lymphocytes from peripheral blood the K_2M_2 hybrid was present, too. Cottreau et al.[122] studied pyruvate kinase in 62 childhood cancers by electrofocusing. It was found that in neuroblastomas a close relationship exists between the differentiation level of tumors and the isozyme composition. The same phenomenon was observed in rhabdomyosarcomas and teratomas. The authors suggested that the isozyme pattern found in the malignant tissues is not similar to their normal fetal counterpart.[122] Beemer et al.[123] studied the isozyme composition of pyruvate kinase in normal retina and retinoblastomas. In fetal retina, five different forms of pyruvate kinase could be detected by electrophoresis. In adult retina the K_4 isozyme is almost absent, while in retinoblastoma the M_4 isozyme is completely absent. Alanine inhibition of pyruvate kinase was in agreement with the electrophoretic pattern. Lopez-Alarcon et al.[124] have measured the sensitivity to ATP of the pyruvate kinase activity of homogenates of human mammary carcinomas of different degrees of histological differentiation. The inhibition by ATP is significantly lower in well-differentiated carcinomas than in poorly differentiated carcinomas. These results were explained on the basis of the isoenzymatic shift from M to K isoenzyme and it was proposed that pyruvate kinase sensitivity to ATP could represent a quantitative method in determining tumor malignancy.[124]

The levels of multimolecular forms of pyruvate kinase present in 4 normal human breast specimens, 9 benign tissues, and 13 malignant breast carcinomas were determined by Ibsen et al.[125] A correlation between specific activity and malignancy was found: malignant tissues showed a much higher specific activity than normal tissue. Both normal and neoplastic tissues primarily express the K isozyme, but the latter show a greater tendency to express more of a subunit form with a pI value similar to that of the M isozyme. This form can, however, be distinguished from that of the M-type by its kinetic properties. Since it is more susceptible to inhibition than is the K-type, increased expression of this form in neoplastic tissues would also explain the increased sensitivity to ATP or alanine inhibition.[125]

V. L-TYPE PYRUVATE KINASE DEFICIENCY

Hereditary erythrocyte pyruvate kinase deficiency is a well-known cause of hemolytic anemia. Since its original description in 1961 by Valentine et al.[126] more than 200 cases of L-type pyruvate kinase deficiency have been reported.[127-130] There is a remarkable variation in the severity of the clinical disease. The spectrum ranges from severe neonatal jaundice, requiring exchange or multiple transfusions at one extreme, to a fully compensated hemolytic process in healthy adults on the other.[127-130] The disease comprises also a heterogeneous group of both quantitative and qualitative pyruvate kinase abnormalities. The majority of the cases showed a decreased affinity for the substrate phospho-enol-pyruvate, although also normal or increased affinities for phospho-enol-pyruvate have been reported.[131-135] In general, no relationship between measurable activity and severity of hemolysis can be demonstrated in pyruvate kinase deficiency.[136] In addition no real relationship is found between abnormalities at the molecular level of pyruvate kinase deficiency and severity of hemolysis as was reported by Zanella et al.,[137] Vives-Corrons et al.,[138] and Elder et al.[139]

By means of recommended methods for the characterization of red cell pyruvate kinase variants, it is possible to distinguish different groups of variants, based on the $R \rightleftharpoons T$ model of Monod.[140]

The first group of variants is characterized by a decreased affinity for the substrate phospho-enol-pyruvate. Patients have been described with pyruvate kinases, characterized by high $K_{0.5}$ for phospho-enol-pyruvate, strongly abnormal utilization of substrate analogues, low fructose-1,6-diphosphate activation, increased ATP and alanine inhibition, and abnormal electrophoretic pattern. However, also patients have been reported with a pyruvate kinase characterized by a high $K_{0.5}$ for phospho-enol-pyruvate but normal activation by fructose-1,6-diphosphate and almost normal substrate analog utilization.[138,141]

The second group of variants is characterized by an increase of the affinity for the substrate phospho-enol-pyruvate; this means in the Monod model a shift in the equilibrium to the R-form.[141] The third group of variants is characterized by normal kinetics, but extreme lability of the enzyme protein, resulting in very low residual enzyme activity.[138,141]

The variants belonging to the third group usually show moderate clinical expression. Despite a very low pyruvate kinase activity the erythrocytes are able to perform a normal glycolytic flux.[141] For the third group of variants, splenectomy can be expected to improve the hematological condition,[138] because young cells have higher pyruvate kinase activity. In contrast, splenectomy in the two other groups of variants may not be successful. After splenectomy the pyruvate kinase activity in the erythrocytes will not change because the kinetic abnormality is still present. It may be important to determine the glycolytic flux of erythrocytes of patients belonging to the first two groups of variants. It is possible that a correlation may be found between the amount of glycolytic flux and the severity of hemolysis.

Van Berkel et al.[142,143] and Sprengers et al.[144] showed that pyruvate kinase can be converted into an oxidized form by incubation with oxidized glutathione. The oxidized enzyme can be reduced again by incubation with mercaptoethanol or with reduced glutathione. The oxidized enzyme shows a lower affinity for the substrate phospho-enol-pyruvate and for the allosteric effector fructose-1,6-diphosphate. The thermolability of the oxidized enzyme is markedly increased. These results indicate that abnormal pyruvate kinases may be induced by an alteration of the redox state in the erythrocyte.[143,144]

REFERENCES

1. **Ibsen, K. H.,** Interrelationships and functions of the pyruvate kinase isozymes and their variant forms: a review, *Cancer Res.*, 37, 341, 1977.
2. **Ibsen, K. H. and Trippet, P.,** A comparison of kinetic parameters obtained with three non-interconvertible isozymes of rat pyruvate kinase, *Arch. Biochem. Biophys.*, 156, 730, 1973.
3. **Van Berkel, Th. J. C.,** Pyruvate Kinase Isoenzymes, Thesis, Erasmus University, Rotterdam, The Netherlands, 1974.
4. **Van Berkel, Th. J. C., Koster, J. F., and Hülsman, W. C.,** Distribution of L- and M-type pyruvate kinase between parenchymal and Kupffer cells of rat liver, *Biochim. Biophys. Acta*, 276, 425, 1972.
5. **Cardenas, J. M., Dyson, R. D., and Strandholm, J. J.,** Pyruvate kinase in adult and fetal tissues of chicken, *Biochemistry*, 14, 2242, 1975.
6. **Tolle, S. W., Dyson, R. D., Newburgh, R. W., and Cardenas, J. M.,** Pyruvate kinase isozymes in neurons, glia, neuroblastoma and glioblastoma, *J. Neurochem.*, 27, 1355, 1976.
7. **Strandholm, J. J., Dyson, R. D., and Cardenas, J. M.,** Bovine pyruvate kinase isozymes and hybrid isozymes. Electrophoretic studies and tissue distribution, *Arch. Biochem. Biophys.*, 173, 125, 1976.
8. **Farina, F. A., Shatton, J. B., Morris, H. P., and Weinhouse, S.,** Isozymes of pyruvate kinase in liver and hepatomas of the rat, *Cancer Res.*, 34, 1439, 1974.
9. **Farron, F., Hsu, H. H. F., and Knox, W. E.,** Fetal-type isoenzymes in hepatic and non-hepatic rat tumors, *Cancer Res.*, 32, 302, 1972.
10. **Kutzbach, C., Bischofberger, H., Hess, B., and Zimmermann-Telschow, H.,** Pyruvate kinase from pig liver, *Z. Physiol. Chem.*, 354, 1473, 1973.
11. **Osterman, J., Fritz, P. J., and Suntch, T.,** Pyruvate kinase isozymes from rat tissues. Developmental studies, *J. Biol. Chem.*, 148, 1011, 1973.
12. **Ostermann, J. and Fritz, P. H.,** Pyruvate kinase isozymes. A comparative study in tissue of various mammalian species, *Comp. Biochem. Physiol.*, 44B, 1077, 1973.
13. **Tanaka, T., Harano, Y., Sue, F., and Morimura, H.,** Crystallization, characterization and metabolic regulation of two types of pyruvate kinase isolated from rat tissues, *J. Biochem.*, 62, 71, 1967.
14. **Imamura, K. and Tanaka, T.,** Multimolecular forms of pyruvate kinase from rat and other mammalian tissues. I. Electrophoretic studies, *J. Biochem.*, 71, 1043, 1972.
15. **Imamura, K., Taniuchi, K., and Tanaka, T.,** Multimolecular forms of pyruvate kinase. II. Purification of M_2-type pyruvate kinase from Yoshida Ascites hepatoma 130 cells and comparative studies on the enzymological and immunological properties of the three types of pyruvate kinase, L, M_1 and M_2, *J. Biochem.*, 71, 1001, 1972.
16. **Harkins, R. N., Black, J. A., and Rittenberg, M. B.,** M_2 isozyme of pyruvate kinase from human kidney as the product of a separate gene: its purification and characterization, *Biochemistry*, 16, 3831, 1977.
17. **Harkins, R. N., Black, J. A., and Rittenberg, M. B.,** Purification and characterization of human muscle pyruvate kinase, *Can. J. Biochem.*, 55, 301, 1977.
18. **Whittell, N. M., Prabhahararao, D. O. K., and Holmes, R. S. A.,** Comparative electrophoretic analysis of mammalian pyruvate kinase isozymes, *Comp. Biochem. Physiol.*, 46, 71, 1973.
19. **Bigley, R. H., Stenzel, P., Jones, R. T., Campos, J. O., and Koler, R. D.,** Tissue distribution of human pyruvate kinase isozymes, *Enzymol. Biol. Clin.*, 9, 10, 1968.

20. **Middleton, M. C. and Walker, D. G.,** Comparison of the properties of two forms of pyruvate kinase in rat liver and determination of their separate activities during development, *Biochem. J.*, 127, 721, 1972.
21. **Carbonell, J., Feliu, J. E., Marco, R., and Sols, A.,** Pyruvate kinase classes of regulatory isoenzymes in mammalian tissues, *Eur. J. Biochem.*, 37, 148, 1973.
22. **Marie, J., Kahn, A., and Boivin, P.,** L-type pyruvate kinase isozymes in man. M-type isozymes in adult and foetal tissues, electrofocusing and immunological studies, *Hum. Genet.*, 31, 35, 1976.
23. **Marie, J., Kahn, A., and Boivin, P.,** L-type pyruvate kinase from human liver. Purification by double affinity elution, electrofocusing and immunological studies, *Biochim. Biophys. Acta*, 438, 393, 1976.
24. **Kahn, A., Marie, J., and Boivin, P.,** Pyruvate kinase isozymes in man. II. L-type and erythrocyte-type isozymes. Electrofocusing and immunological studies, *Hum. Genet.*, 33, 35, 1976.
25. **Kahn, A., Marie, J., Garreau, H., and Sprengers, E. D.,** The genetic system of the L-type pyruvate kinase forms in man. Subunit structure, interrelation and kinetic characteristics of the pyruvate kinase enzymes from erythrocytes and liver, *Biochim. Biophys. Acta*, 523, 59, 1978.
26. **Sprengers, E. D. and Staal, G. E. J.,** Functional changes associated with the sequential transformation of L'_4 into L_4 pyruvate kinase, *Biochim. Biophys. Acta*, 570, 259, 1979.
27. **Badwey, J. A. and Westhead, E. W.,** Hysteretic response of human erythrocyte pyruvate kinase to phospho-enol-pyruvate, *J. Biol. Chem.*, 251, 5600, 1976.
28. **Marie, J., Tichonicky, L., Dreyfus, J. C., and Kahn, A.,** Endogenous, cyclic 3'-5'-AMP dependent phosphorylation of human red cell pyruvate kinase, *Biochem. Biophys. Res. Commun.*, 87, 862, 1979.
29. **Marie, J. and Kahn, A.,** Proteolytic processing of human L-type pyruvate kinase increases its ability to be phosphorylated, *Biochem. Biophys. Res. Commun.*, 94, 1387, 1980.
30. **Lagrange, J. L., Marie, J., Cottreau, D., Fischer, S., and Kahn, A.,** Endogenous phosphorylation of soluble enzymes in human red cells, *Biochim. Biophys. Acta*, 612, 213, 1980.
31. **Bigley, R. H. and Koler, R. D.,** Liver pyruvate kinase (PK) isozymes in a PK-deficient patient, *Ann. Hum. Genet.*, 312, 383, 1986.
32. **Kahn, A., Marie, J., Galand, C., and Boivin, P.,** Chronic haemolytic anemia in two patients heterozygous for erythrocyte pyruvate kinase deficiency. Electrofocusing and immunological studies of erythrocyte and liver pyruvate kinase, *Scand. J. Haematol.*, 16, 250, 1976.
33. **Nakashima, K., Miwa, S., and Fujii, H.,** Characterization of pyruvate kinase from the liver of a patient with aberrant erythrocyte pyruvate kinase, PK Nagashima, *J. Lab. Clin. Med.*, 90, 1012, 1977.
34. **Minakami, W., Suzuki, C., Saito, T., and Yoshikawa, W.,** Studies on erythrocyte glycolysis. I. Determination of the glycolytic intermediates in human erythrocytes, *J. Biochem. (Tokyo)*, 58, 543, 1965.
35. **Staal, G. E. J., Rijksen, G., Vlug, A. M. C., Vromen-van den Bos, B., Akkerman, J. W. N., Gorter, G., Dierick, J., and Petermans, M.,** Extreme deficiency of L-type pyruvate kinase with moderate clinical expression, *Clin. Chim. Acta*, 118, 241, 1982.
36. **Marie, J., Simon, M. P., Dreyfus, J. C., and Kahn, A.,** One gene, but two messenger RNA's encode liver L and red cell L'pyruvate kinase subunits, *Nature (London)*, 292, 70, 1981.
37. **Saheki, S., Saheki, K., and Tanaka, T.,** Peptide structures of pyruvate kinase isozymes. I. Comparison of the four pyruvate kinase isozymes of the rat, *Biochim. Biophys. Acta*, 704, 484, 1982.
38. **Ibsen, K. H., Chiu, R. H. C., Park, H. R., Sanders, D. A., Roy, S., Garratt, K. N., and Mueller, M. K.,** Purification and properties of mouse pyruvate kinases K and M and of a modified K subunit, *Biochemistry*, 20, 1497, 1981.
39. **Muroya, N., Nago, Y., Miyasaki, K., Nishikawa, K., and Horio, T.,** Pyruvate kinase isozymes in various tissues of rat and increase in spleen type pyruvate kinase in liver by injecting chromatins from spleen and tumor, *J. Biochem.*, 79, 203, 1976.
40. **Noguchi, T. and Tanaka, T.,** The M_1 and M_2 subunits of rat pyruvate kinase are encoded by different messenger RNA's, *J. Biol. Chem.*, 257, 1110, 1982.
41. **Saheki, S., Saheki, K., Tanaka, T., and Tanaka, T.,** Peptide structures of pyruvate kinase isozymes. II. Origins of type M_1 and M_2 isozymes suggested from species-variations in their peptide maps, *Biochim. Biophys. Acta*, 704, 494, 1982.
42. **Levin, M. J., Daegelen, D., Meienhofer, M. C., Dreÿfus, J. C., and Kahn, A.,** Two different species of messenger RNA's specify synthesis of M_1 and M_2 pyruvate kinase subunits, *Biochim. Biophys. Acta*, 699, 77, 1982.
43. **Hall, E. R. and Cottam, G. L.,** Isozymes of pyruvate kinase in vertebrates: their physical, chemical, kinetic and immunological properties, *Int. J. Biochem.*, 9, 785, 1978.
44. **Carminatti, H., Jiménez de Asua, L., Recondo, E., Passeron, S., and Rozengurt, F.,** Some kinetic properties of liver pyruvate kinase (type-L), *J. Biol. Chem.*, 243, 3051, 1968.
45. **Rozengurt, E., Jiménez de Asua, L., and Carminatti, H.,** Some kinetic properties of liver pyruvate kinase (type-L). II. Effect of pH on its allosteric behaviour, *J. Biol. Chem.*, 244, 3142, 1969.
46. **Seubert, W. and Shoner, W.,** The regulation of pyruvate kinase, in *Current Topics in Cellular Regulation*, Vol. 3, Horecker, B. L. and Stadtman, E. R., Eds., Academic Press, New York, 1971, 237.

47. **Koster, J. F. and Hülsmann, W. C.,** The influence of inorganic phosphate and phosphorylated hexoses on the activity of pyruvate kinase, *Arch. Biochem. Biophys.*, 141, 98, 1970.
48. **Koster, J. F., Slee, R. G., Staal, G. E. J., and Van Berkel, Th. J. C.,** The influence of glucose-1,6-diphosphate on the enzymatic activity of pyruvate kinase, *Biochim. Biophys. Acta*, 258, 763, 1972.
49. **Van Berkel, Th. J. C., Koster, J. F., Kruyt, J. K., and Hülsmann, W. C.,** On the regulation and allosteric model of L-type pyruvate kinase from rat liver, *Biochim. Biophys. Acta*, 370, 450, 1974.
50. **Beitner, R.,** The role of glucose-1,6-bisphosphate in the regulation of carbohydrate metabolism in muscle, *Trends Biochem. Sci.*, 4, 228, 1979.
51. **Hess, H. G. and Van Schaftingen, E.,** Fructose-2,6-biphosphate 2 years after discovery, *Biochem. J.*, 206, 1, 1982.
52. **Pilkis, S. J., El-Maghrabi, M. R., McGrane, M., Pilkis, J., Fox, E., and Claus, T. H.,** Fructose-2,6-biphosphate: a mediator of hormone action at the fructose-6-phosphate/fructose-1,6-biphosphate substrate cycle, *Mol. Cell. Endocrinol.*, 25, 245, 1982.
53. **Staal, G. E. J., Koster, J. F., Kamp, H., Van Milligen-Boersma, L., and Veeger, C.,** Human erythrocyte pyruvate kinase, its purification and some properties, *Biochim. Biophys. Acta*, 227, 86, 1971.
54. **Rozengurt, E., Jiménez de Asua, L., and Carminatti, H.,** Allosteric inhibition of muscle pyruvate kinase by phenyl alanine, *FEBS Lett.*, 11, 284, 1970.
55. **Van Berkel, Th. J. C. and Koster J. F.,** M-type pyruvate kinase of leukocytes: an allosteric enzyme, *Biochim. Biophys. Acta*, 293, 134, 1973.
56. **Jiménez de Asua, L., Rozengurt, E., DeValle, J. J. and Carminatti, H.,** Some kinetic differences between the M isoenzymes of pyruvate kinase from liver and muscle, *Biochim. Biophys. Acta*, 235, 326, 1971.
57. **Monod, J., Wyman, J., and Changeux, J. P.,** On the nature of allosteric transitions: a plausible model, *J. Mol. Biol.*, 12, 88, 1965.
58. **Van Berkel, Th. J. C., Koster, J. F., and Hülsmann, W. C.,** Some kinetic properties of the allosteric M-type pyruvate kinase from rat liver; influence of pH and the nature of amino acid inhibition, *Biochim. Biophys. Acta*, 321, 171, 1973.
59. **Pogson, C. J.,** Adipose-tissue pyruvate kinase. Properties and interconversion of two active forms, *Biochem. J.*, 110, 67, 1968.
60. **Walker, P. R. and Potter, V. R.,** Allosteric properties of a pyruvate kinase isozyme from rat liver cells in culture, *J. Biol. Chem.*, 248, 4620, 1973.
61. **Van Berkel, Th. J. C.,** Some kinetic properties of M_2-type pyruvate kinase from rat liver at physiological Mg^{2+} concentration, *Biochim. Biophys. Acta*, 370, 140, 1974.
62. **Van Berkel, Th. J. C., de Jonge, H. R., Koster, J. F., and Hülsmann, W. C.,** Kinetic evidence for the presence of two forms of M_2-type pyruvate kinase in rat small intestine, *Biochem. Biophys. Res. Commun.*, 60, 298, 1974.
63. **Osterman, J. and Fritz, P. J.,** Pyruvate kinase isozymes from rat intestinal mucosa. Characterization and the effect of fasting and refeeding, *Biochemistry*, 13, 1731, 1973.
64. **Ljungström, O., Hjelmquist, G., and Engström, L.,** Phosphorylation of purified rat liver pyruvate kinase by cyclic 3′,5′,-AMP-stimulated protein kinase, *Biochim. Biophys. Acta*, 358, 289, 1974.
65. **Titanji, V. P. K., Zetterquist, D., and Engström, L.,** Regulation in vitro of rat liver pyruvate kinase by phosphorylation-dephosphorylation reactions, catalyzed by cyclic-AMP dependent protein kinases and a histone phosphatase, *Biochim. Biophys. Acta*, 422, 98, 1976.
66. **Berglund, L., Ljungström, O., and Engström, L.,** Studies on the cyclic 3′,5′,-AMP-stimulated pig liver protein kinase reaction with pyruvate kinase as substrate, *J. Biol. Chem.*, 252, 613, 1977.
67. **Marie, J., Tichonicky, L., Dreyfus C., and Kahn, A.,** Endogenous, cyclic 3′,5′-AMP-dependent phosphorylation of human red cell pyruvate kinase, *Biochem. Biophys. Res. Commun.*, 87, 862, 1979.
68. **Ekman, P., Dahlquist, U., Humble, E., and Engström, L.,** Comparative kinetic studies on the L-type pyruvate kinase from rat liver and the enzyme phosphorylated by cyclic 3′,5′,-AMP-stimulated protein kinase, *Biochim. Biophys. Acta*, 429, 374, 1976.
69. **Claus, T. H., El-Maghrabi, M. R., and Pilkis, S. J.,** Modulation of the phosphorylation state of rat liver pyruvate kinase by allosteric effectors and insulin, *J. Biol. Chem.*, 254, 7855, 1979.
70. **Pilkis, W. J., Pilkis, J., and Claus, T. H.,** The effect of fructose diphosphate and phosphoenol-pyruvate on cyclic AMP-mediated inactivation of rat hepatic pyruvate kinase, *Biochem. Biophys. Res. Commun.*, 81, 139, 1978.
71. **El-Maghrabi, M. R., Claus, T. H., McGrave, M. M., and Pilkis, S. J.,** Influence of phosphorylation on the interaction of effectors with rat liver pyruvate kinase, *J. Biol. Chem.*, 257, 233, 1982.
72. **Riou, J. P., Claus, T. H., and Pilkis, S. J.,** Stimulation by glucagon of in vivo phosphorylation of rat hepatic pyruvate kinase, *J. Biol. Chem.*, 251, 3756, 1976.

73. **Blair, J. B., Cimbala, M. A., Foster, J. L., and Morgan, R. A.,** Hepatic pyruvate kinase. Regulation by glucagon, cyclic adenosine, 3',5',-monophosphate, and insulin in the perfused rat liver, *J. Biol. Chem.*, 251, 3756, 1976.
74. **Clark, M. G.,** Glucagon regulation of pyruvate kinase in relation to phosphoenol-pyruvate substrate cycling in isolated rat hepatocytes, *Biochem. Biophys. Res. Commun.*, 68, 120, 1976.
75. **Ljungström, O. and Ekman, P.,** Glucagon-induced phosphorylation of pyruvate kinase (type L) in rat liver slices, *Biochem. Biophys. Res. Commun.*, 78, 1147, 1977.
76. **Felin, J. E., Hue, L., and Hers, H. G.,** Hormonal control of pyruvate kinase activity and of gluconeogenesis in isolated hepatocytes, *Proc. Natl. Acad. Sci. U.S.A.*, 73, 2762, 1976.
77. **Johnson, M. L. and Veneziale, C. M.,** Hormonal regulation of liver pyruvate kinase concentration and activity, *Biochemistry*, 19, 2191, 1980.
78. **Van Berkel, Th. J. C., Kruyt, J. K., Koster, J. F., and Hülsmann, W. C.,** Cyclic nucleotide-, pyruvate-, and hormone-induced changes in pyruvate kinase activity in isolated rat hepatocytes, *Biochem. Biophys. Res. Commun.*, 72, 917, 1976.
79. **Van den Berg, G. B., Van Berkel, Th. J. C., and Koster, J. F.,** Identification of L-type pyruvate kinase as a major phosphorylation site of endogenous cyclic AMP-dependent protein kinase in rat liver soluble fraction, *FEBS Lett.*, 101, 289, 1979.
80. **Van Berkel, Th. J. C., Kruyt, J. K., and Koster, J. F.,** Hormone-induced changes in pyruvate kinase. Effects of glucagon and starvation, *Eur. J. Biochem.*, 81, 423, 1977.
81. **Van Berkel, Th. J. C., Kruyt, J. K., van den Berg, G. B., and Koster, J. F.,** Difference in the effect of glucagon and starvation upon L-type pyruvate kinase from rat liver, *Eur. J. Biochem.*, 92, 553, 1978.
82. **Kohl, E. A. and Cottam, G. L.,** Activation in liver pyruvate kinase protein and catalytic activity upon starvation and refeeding, *Arch. Biochem. Biophys.*, 176, 671, 1976.
83. **Blair, J. B., Cimbala, M. A., and James, M. E.,** Hepatic pyruvate kinase. Quantitative measurements of phosphorylation in vitro and in the isolated rat hepatocyte, *J. Biol. Chem.*, 257, 7595, 1982.
84. **Bergström, G., Ekman, P., Humble, E., and Engström, L.,** Proteolytic modification of pig and rat liver pyruvate kinase type L including phosphorylatable site, *Biochim. Biophys. Acta*, 532, 259, 1978.
85. **Marie, J., Buc, H., Simon, M. P., and Kahn, A.,** Phosphorylation of human erythrocyte pyruvate kinase by soluble cyclic– AMP-dependent protein kinases, *Eur. J. Biochem.*, 108, 251, 1980.
86. **Weber, G.,** Enzymology of cancer cells. I, *N.Engl. J. Med.*, 296, 486, 1977.
87. **Weber, G.,** Enzymology of cancer cells. II, *N. Engl. J. Med.*, 296, 541, 1977.
88. **Weinhouse, S.,** Metabolism and isozyme alterations in experimental hepatomas, *Fed. Proc. Fed. Am. Soc. Exp. Biol.*, 32, 2162, 1973.
89. **Schapira, F.,** Isozymes and cancer, *Adv. Cancer Res.*, 18, 77, 1973.
90. **Weinhouse, S.,** New dimensions in the biology of cancer, *Cancer*, 45, 2975, 77.
91. **Weinhouse, S.,** What are isozymes telling us about gene regulation in cancer?, *J. Natl. Cancer Inst.*, 68, 343, 1982.
92. **Schapira, F.,** Resurgence of fetal isozymes in cancer: study of aldolase, pyruvate kinase, lactic dehydrogenase and β-hexosaminidase, in *Isozymes: Current Topics in Biological and Medical Research*, Liss, A. R., Ed., New York, 1981, 27.
93. **Ibsen, H. I. and Fishman, W. H.,** Developmental gene expression in cancer, *Biochim. Biophys. Acta*, 560, 243, 1979.
94. **Tanaka, T., Harano, Y., Morimura, H., and Mori, R.,** Evidence for the presence of two types of pyruvate kinase in rat liver, *Biochem. Res. Commun.*, 21, 55, 1965.
95. **Taylor, C. B., Morris, H. P., and Weber, G.,** A comparison of the properties of pyruvate kinase from hepatoma 3924-A, normal liver and muscle, *Life Sci.*, 8, 635, 1969.
96. **Schapira, F. and Gregori, C.,** Pyruvate kinase de l'hépatome, du placenta et du foie foetal de rat, *C. R.*, 272, 1169, 1971.
97. **Weinhouse, S.,** Enzymatic and metabolic alterations in experimental hepatomas, in *Differentiation and Control of Malignancy of Tumor Cells*, Nakahara, W., Ono, T., Sugimura, T., and Sugano, H., Eds., University Park Press, Baltimore, 1977, 187.
98. **Weinhouse, S., Shatton, J. B., Criss, W. E., Farina, F. A., and Morris, H. P.,** Isozymes in relation to differentiation in transplantable rat hepatomas, *Gann Monogr.*, 13, 1, 1972.
99. **Weinhouse, S., Shatton, J. B., Criss, W. E., and Morris, H. P.,** Molecular forms of enzymes in cancer, *Biochimie*, 54, 685, 1972.
100. **Tanaka, T., Yanagi, S., Imamura, K., Kashiwagi, A., and Ito, N.,** Deviations in patterns of multimolecular forms of pyruvate kinase in tumor cells and in the liver of tumor-bearing animals, in *Differentiation and Control of Malignancy of Tumor Cells*, Nakahara, W., Ono, T., Sugimura, T., and Sugano, H., Eds., University Park Press, Baltimore, 1974, 221.

101. **Yanagi, S., Makiura, S., Arai, M., Matsumura, K., Hirao, K., Ito, N., and Tanaka, T.,** Isozyme patterns of pyruvate kinase in various primary liver tumors induced during the process of hepatocarcinogenesis, *Cancer Res.*, 34, 2283, 1974.
102. **Weber, P. R. and Potter, V. R.,** Isozyme studies on adult, regenerating, precancerous and developing liver in relation to findings in hepatomas, *Adv. Enzyme Regul.*, 10, 339, 1972.
103. **Guguen, C., Gregori, C., and Schapira, F.,** Modification of pyruvate kinase isozymes in prolonged primary cultures of adult rat hepatocytes, *Biochimie*, 57, 1065, 1975.
104. **Walker, P. R., Bonney, R. J., Becher, J. E., and Potter, V. R.,** Pyruvate kinase, hexokinase and aldolase in liver during development, *Life Sci.*, 15, 1507, 1974.
105. **Ibsen, K. H., Basabe, J. R., and Lopez, T. P.,** Extraction of a factor from Ehrlich Ascites tumor cells that increases the activity of the fetal isozyme of pyruvate kinase in mouse liver, *Cancer Res.*, 35, 180, 1975.
106. **Nakamura, T., Hosoi, K., Nishikawa, K., and Horio, T.,** Effect of injection of chromatins from Rhodamine sarcoma into rats on pI-isozyme of liver pyruvate kinase, *Gann*, 63, 239, 1972.
107. **Suda, M., Tanaka, T., Sue, F., Harano, Y., and Morimura, H.,** Dedifferentiation of sugar metabolism in the liver of tumor bearing rats, *Gann Monogr.*, 1, 127, 1966.
108. **Suda, M., Tanaka, T., Yanagi, S., Hayashi, S., Imamura, K., and Taniuchi, K.,** Dedifferentiation of enzymes in the liver of tumor-bearing animals, *Gann Monogr.*, 13, 79, 1972.
109. **Suda, M., Tanaka, T., Sue, F., Kuroda, Y., and Morimura, H.,** Rapid increase of pyruvate kinase M-type) and hexokinase in normal rat liver by perfusion of the blood of tumor-bearing rat, *Gann Monogr.*, 4, 103, 1968.
110. **Van Veelen, C. W. M., Verbiest, H., Vlug, A. M. C., Rijksen, G., and Staal, G. E. J.,** Isozymes of pyruvate kinase from human brain, meningiomas and malignant gliomas, *Cancer Res.*, 38, 4681, 1978.
111. **Bennett, M. J., Timperley, W. R., Taylor, C. B., and Hill, A. S.,** Fetal forms of pyruvate kinase isoenzymes in tumours of the human nervous system, *Neuropathol. Appl. Neurobiol.*, 1, 347, 1975.
112. **Van Veelen, C. W. M., Rijksen, G., Vlug, A. M. C., and Staal, G. E. J.,** Correlation between alanine inhibition of pyruvate kinase and composition of K-M hybrids, *Clin. Chim. Acta*, 110, 113, 1981.
113. **Van Veelen, C. W. M., Verbiest, H., Zülch, K. J., van Ketel, B. A., van der Vlist, M. J. M., Vlug, A. M. C., Rijksen, G., and Staal, G. E. J.,** L-α-alanine inhibition of pyruvate kinase from tumors of the human central nervous system, *Cancer Res.*, 39, 4263, 1979.
114. **Van Veelen, C. W. M., Verbiest, H., Zülch, K. J., van Ketel, B. A., van der Vlist, M. J. M., Vlug, A. M. C., Rijksen, G., and Staal, G. E. J.,** L-α-alanine inhibition of pyruvate kinase from tumours of the human central nervous system. A new aid in the treatment of gliomas, in *Modern Neurosurg.*, Brock, M., Ed., Springer-Verlag, Berlin, 1982, 110.
115. **Van Veelen, C. W. M., Verbiest, H., Zülch, K. J., van Ketel, B., van der Vlist, M. J. M., Vlug, A. M. C., Rijksen, G., and Staal, G. E. J.,** Pyruvate kinase in human brain tumours. Its significance in the treatment of gliomas, *Acta Neurochir.*, 61, 145, 1982.
116. **Gosalvez, M., Lopez-Alcarcon, L., Garica-Suarez, S., Montalvo, A., and Weinhouse, S.,** Stimulation of tumor-cell respiration by inhibitors of pyruvate kinase, *Eur. J. Biochem.*, 55, 315, 1975.
117. **Gosalvez, M., Barcia-Suarez, S., and Lopez-Alcarcon, L.,** Metabolic control of glycolysis in normal and tumor permeabilized cells, *Cancer Res.*, 38, 142, 1978.
118. **Farron, F., Hsu, H. H., and Knox, W. E.,** Fetal-type isoenzymes in hepatic and nonhepatic rat tumors, *Cancer Res.*, 32, 302, 1972.
119. **Nagao, Y., Toda, T., Miyazaki, K., and Horio, T.,** Crystallization of spleen-type (M_2-type) pyruvate kinase from Rhodamine sarcoma of rats and its properties, *J. Biochem.*, 82, 1331, 1977.
120. **Kraaijenhagen, R. J., de Gast, G. C., van der Heijden, M. C. M., Streefkerk, M., Gmelig-Meyling, F. H. J., Rijksen, G., and Staal, G. E. J.,** Isozyme distribution of hexokinase, phosphofructokinase and pyruvate kinase in lymphocytes from patients with chronic lymphocytic leukemia, *Clin. Chim. Acta*, 124, 91, 1982.
121. **Kraaijenhagen, R. J., van der Heijden, M. C. M., Streefkerk, M., Rijksen, G., de Gast, G. C., and Staal, G. E. J.,** Isozyme patterns of glycolytic regulator enzymes in T and B lymphoblastic cell lines, compared with T and non-T lymphocytes from peripheral blood, *Hum. Lymph. Diff.*, 1, 241, 1981.
122. **Cottreau, D., Rousseau-Merck, M. F., Nezelof, C., and Kahn, A.,** Pyruvate kinase and phosphofructokinase isozymes in childhood cancers, *Pediatr. Res.*, 16, 199, 1982.
123. **Beemer, F. A., Vlug, A. M. C., Rijksen, G., Hamburg, A., and Staal, G. E. J.,** Characterization of some glycolytic enzymes from human retina and retinoblastoma, *Cancer Res.*, 42, 4228, 1982.
124. **Lopez-Alarcon, L., Ruiz, P., and Gosalvez, M.,** Quantitative determination of the degree of differentiation of mammary tumors by pyruvate kinase kinetic analysis, *Cancer Res.*, 41, 2019, 1981.
125. **Ibsen, K. H., Orlando, R. A., Garratt, K. N., Hernandez, A. M., Giorlando, S., and Nungaray, G.,** Expression of multimolecular forms of pyruvate kinase in normal, benign and malignant human breast tissue, *Cancer Res.*, 42, 888, 1982.

126. **Valentine, W. N., Tanaka, K. R., and Miwa, S.,** Specific erythrocyte glycolytic enzyme defect (pyruvate kinase) in three subjects with congenital non-spherocytic hemolytic anemia, *Trans. Assoc. Am. Phys.*, 74, 100, 1961.
127. **Boivin, P.,** Déficit en pyruvate kinase, in *Enzymopathies*, Boivin, P., Gajdos, A., and Lestradet, H., Eds., Masson ; am Cie, Paris, 1971, 190.
128. **Tanaka, K. R., and Paglia, D. E.,** Pyruvate kinase deficiency, *Semin. Hematol.*, 8, 367, 1971.
129. **Miwa, S.,** Hereditary hemolytic anemia due to erythrocyte enzyme deficiency, *Acta Haematol. Jpn.*, 36, 573, 1973.
130. **Beutler, E.,** Hemolytic anemia in disorder of red cell population, in *Topics in Hematology*, Plenum Press, New York, 1978, 169.
131. **Staal, G. E. J., Koster, J. F., and Van Milligen-Boersma, L.,** Some properties of abnormal red blood cell pyruvate kinase, *Biochim. Biophys. Acta*, 220, 613, 1970.
132. **Blume, K. G., Arnold, H., Löhr, G. W., and Scholtz, G.,** On the molecular basis of pyruvate kinase deficiency, *Biochim. Biophys. Acta*, 370, 601, 1974.
133. **Nakashima, K., Miwa, S., Oda, S., Tanaka, T., Imamura, K., and Nishima, T.,** Electrophoretic and kinetic studies of mutant erythrocyte pyruvate kinases, *Blood*, 43, 537, 1974.
134. **Staal, G. E. J., Geerdink, R. P., Vlug, A. M. C., and Hamelink, M. L.,** Defective erythrocyte pyruvate kinase, *Clin. Chim. Acta*, 68, 11, 1976.
135. **Marie, J., Vives-Corrons, J. L., and Kahn, A.,** Hereditary erythrocyte pyruvate kinase deficiency, molecular and functional studies of four mutant PK variants detected in Spain, *Clin. Chim. Acta*, 81, 153, 1977.
136. **Kahn, A., Kaplan, J. C., and Dreyfus, J. C.,** Advances in hereditary red cell enzyme anomalies, *Hum. Genet.*, 50, 1, 1979.
137. **Zanella, A., Rebulla, P., Izzo, C., Tedesco, F., and Sircha, G.,** Hereditary pyruvate kinase deficiency. Role of the abnormal enzyme in red cell pathophysiology, *Br. J. Haematol.*, 40, 551, 1978.
138. **Vives-Corrons, J. L., Marie, J., Pujades, M. A., and Kahn, A.,** Hereditary erythrocyte pyruvate kinase (PK) deficiency and chronic hemolytic anemia: clinical, genetic and molecular studies in six new Spanish patients, *Hum. Genet.*, 53, 401, 1980.
139. **Elder, G. E., Lappin, T. R. J., Lawson, B. E., and Bridges, J. M.,** Three pyruvate kinase variants with increased affinity for PEP, *Br. J. Haematol.*, 47, 371, 1981.
140. Recommended methods for the characterization of red cell pyruvate kinase variants, *Br. J. Haematol.*, 43, 275, 1981.
141. **Staal, G. E. J., Rijksen, G., Vlug, A. M. C., Vromen-van den Bos, B., Akkerman, J. W. N., Gorter, G., Dierick, J., and Petermans, M.,** Extreme deficiency of L-type pyruvate kinase with moderate clinical expression, *Clin. Chim. Acta*, 118, 241, 1982.
142. **Van Berkel, Th. J. C., Koster, J. F., and Staal, G. E. J.,** On the molecular basis of pyruvate kinase deficiency. I. Primary defect or consequence of increased glutathione disulfide concentration, *Biochim. Biophys Acta*, 321, 496, 1973.
143. **Van Berkel, Th. J. C., Staal, G. E. J., Koster, J. F., and Nijessen, J. G.,** On the molecular basis of pyruvate kinase deficiency. II. Role of thiol groups in pyruvate kinase from pyruvate kinase-deficiency patients, *Biochim. Biophys. Acta*, 334, 361, 1974.
144. **Sprengers, E. D., Van Berkel, Th. J. C., Koster, J. F., and Staal, G. E. J.,** Influence of the redox state of glutathione upon pyruvate kinase in the intact erythrocyte, *Clin. Chim. Acta*, 80, 495, 1977.

INDEX

A

B

C

I

K

P

R

S

T

U

V

W